II
IN HEALTHCARE FACILITIES

医疗导视 II

(美)吉姆·哈丁 编 常文心 译

辽宁科学技术出版社

前言

Jim Harding

Jim Harding, SEGD, is director of Gresham, Smith and Partners' award-winning Environmental Graphic Design group. His vast signage and wayfinding design experience is unique in the breadth of industries and project types it covers. Jim is frequently published in major industry trade journals, and recently authored the Airport Cooperative Research Program's Wayfinding and Signage Guidelines for Airport Terminals and Landsides, a first-of-its-kind guidebook presenting airport operators across America with accepted best practices for terminal and roadway signage. He is a member of the Society for Environmental Graphic Design and has been honored more than a dozen times for his planning and design of signage programs at corporate offices, universities, hospitals, airports and other major developments from coast to coast.

吉姆·哈丁（环境图形设计协会成员）是 GS&P 设计公司的环境图形设计团队的总监，具有丰富的导视设计经验，涉及广泛的产业和项目类型。吉姆经常在业内主要期刊上发表作品，最近还编写了"机场合作研究项目"的《机场航站楼及公共场所导视设计指南》，该书是首本呈现了全美机场运营中优秀的航站楼及路面引导标识的指导手册。作为环境图形设计协会成员，他的导视设计作品涉及国内外各种企业办公、大学、医院、机场等大型开发项目，获得了十几项国内外大奖。

Congratulations to everyone who contributed to the content in this book! You are among a select group of design professionals in the world.

It's been an exciting road since then, and I'm very proud to be among the small group of people around the world who call themselves environmental graphic designers. We are fortunate to have found a career that lets us unite our passions for design and communication. It's an exciting field that has taught me the value of teamwork and inspiration.

For 29 years, I've led the environmental graphic design team within the international architecture/engineering firm Gresham, Smith and Partners. We're a small but robust group, and many of our best ideas are generated by the reciprocal energy of our team. Each of us understand that we are an important piece of a larger puzzle, contributing unique design philosophies to our projects. Our group believes that inspiration can come from anywhere at any time, so we work hard to capture the sentiment of the things that inspire us and share them with our teammates as often as possible. Someone might email a photo of a sculpture they loved while on a trip or make a simple statement during a meeting that triggers an idea. We track these inspirations so we can to tap into them whenever we need. Collaboration drives us, and idea-sharing is essential to our clients' success.

We also recognise the importance of research in each of our projects. Sufficient study and analysis is crucial for establishing realistic expectations of what environmental graphic design can and can't do. While good signage can't always overcome architectural barriers or non-intuitive wayfinding environments, it can still help achieve a high wayfinding success rate and a strong sense of place for the majority of users. Each project has its own unique environment, objectives and challenges, but research is the common foundation that allows us to establish a baseline from which we can measure the impact and success of the design. Our research begins with data collection: listening to and understanding our clients and their customers, then thinking through existing conditions, identifying obstacles and establishing objectives before arriving at a solution based on reliable data.

Visual wayfinding, the most basic and straightforward navigational tool, encompasses all static signage. It's the workhorse of the wayfinding world; it does the heavy lifting. The success of visual wayfinding is tied strongly to intuitive architecture. Research has shown time and time again that good wayfinding begins with thoughtful architectural design. Visual wayfinding is also the most effective method of presenting and reinforcing a client's brand with clear and cohesive visual elements. A primary challenge of visual wayfinding is balancing form and function. The most important part of our job is turning ideas and inspiration into something real that does a job and does it well. Environmental graphic design must be a hybrid of art and communication. EGD projects need to be both aesthetically pleasing and useful, packaging and presenting clear and concise information to users. Context plays a tremendous role in the success of wayfinding, influencing how a message is comprehended and how decisions are made. It's strongly related to customer expectations, in that information should be provided in the right way at the right place at the right time. Visitors walking down a long hall at a hospital need to be reassured by signage that they're headed in the correct direction. Failure to meet customer expectations can result in people getting lost or feeling a lack of certainty as they navigate through a space.

Verbal wayfinding is another piece of the puzzle. For users who need further instruction on how to reach their destination, or for those who may not speak the language displayed on signage, an information desk offering verbal assistance is very important. Desk attendants should be able to present information in an educated, consistent and objective manner, so users are less likely to become confused when listening to directions. While visual wayfinding is usually successful for around 90% of users, the verbal component is key in assisting the remaining 10%. At a high-traffic hospital, this could translate into thousands of patients and visitors per year.

Virtual wayfinding encompasses dynamic, non-static navigational tools; in essence, digital tools: computerised displays with directional information, interactive directories with foreign language assistance, "smart garage" signage with real-time parking-spot counts, smartphone apps that guide users through a facility, and plenty of other technology. Virtual wayfinding offerings are constantly improving and they will continue to be a crucial component of a comprehensive wayfinding system into the future.

Each of the 3Vs provide and reinforce the same information; it's just presented and accessed in different ways since people process information differently. Where visual signage might be insufficient for one user, verbal wayfinding can fill in the gaps; where virtual information isn't enough for another user, architectural cues guide him; and so forth. By communicating with each other and ensuring there are no contradictions or missing puzzle pieces, the various types of wayfinding are able to consistently present information across all three platforms. Applying a holistic approach so the 3Vs can work in tandem is critical. Using this strategy enables clients to reach the greatest percentage of their customer base, which means that more users can quickly and easily find their destination and comprehend a brand. By tapping into teamwork, creative inspiration and best practices for information delivery, we can elevate the design process and achieve successful outcomes for our clients.

The EGD profession plays a vital role in how people experience a space or place; therefore, every project we touch impacts people's quality of life in some form or fashion. The projects featured here focus on the healthcare market, where customers are likely not at their best. Being lost or confused only adds to their burden. Every design professional represented in this book has both the honor and responsibility to provide clients and, more importantly, their customers, with creative solutions that solve real problems. The opportunity to see how our design efforts are making a real difference every day is the true payoff for all of our hard work. The added benefit of being recognised in this book for a job well done is a welcome validation. Congratulations once again to everyone that contributed to content that follows!

Jim Harding
Gresham, Smith and Partners, USA

祝贺所有对本书内容做出贡献的各位！你们已经跻身于全球设计精英的行列了。

这是一条令人激动的道路，我很自豪能跻身于少数环境图形设计师的行列。我们很幸运，能够找到一个统一了设计与交流的事业。这是一个激动人心的领域，它教会了我团队合作与灵感的价值。

29年来，我一直领导着GS&P设计公司的环境图形设计团队。我们的团队虽小，却很能干，许多最佳概念都来自于团队的互助能量。我们每个人都了解自己是一大幅拼图中很重要的一块，不断为我们的项目贡献独特的设计理念。我们的团队坚信灵感可以来自任何时间、任何地点，所以我们努力捕捉每件事的感悟，并尽量经常与团队成员分享。某人可能发一张喜欢的雕塑的照片，然后在会议上做出简单的评论。我们不断追踪这些灵感，然后在需要的时候对其进行挖掘。合作是我们的驱动力，而思想共享是我们成功的关键。我们还认识到了对每个项目进行调研的重要性。充足的研究和分析对预计环境图形设计的预期效果是至关重要的。虽然好的导视设计不一定能克服建筑障碍或非直观导航环境，但是它仍有助于提升导航的成功率并为大多数用户打造强烈的场所感。每个项目都有其特定的环境、目标和挑战，但是研究是我们建立基准线的共同基础。我们通过基准线来衡量设计的影响力和成功与否。我们的调研始于数据收集：倾听并理解客户和消费者的需求，然后仔细考虑现有条件，识别障碍，在提出解决方案前先根据可靠数据建立起目标。

环境图形设计不仅是标识设计，它要求根据交流信息进行整体研究，帮助人们以最佳方式到达目的地。它有三个基本元素：视觉、语言和虚拟（visual, verbal, virtual, 即3V元素）。各种类型的寻路设计都有其独特的价值，它们融合起来，共同形成一个全面运作的综合系统。

视觉寻路元素是最基本、最直接的导航工具，包含所有静态标识。它是寻路世界中的主力。视觉寻路设计的成功与直观建筑紧密相连。研究反复表明，良好的寻路系统始于合适的建筑设计。视觉寻路设计也是最有效的导航方式，它能通过鲜明、有凝聚力的视觉元素呈现并强化客户的品牌形象。视觉寻路设计的主要挑战是如何平衡形式与功能。在我们工作中，最重要的部分就是将想法与灵感转化为真实而有用的东西。环境图形设计必须综合艺术和交流。环境图形设计项目既要美观实用，又必须为用途打包和呈现清晰、准确的信息。环境在成功的寻路设计中扮演了极为重要的角色，它能影响信息的理解和决策的制定。它与客户的期望效果紧密相连，即信息应当以正确的方式出现在正确的时间、正确的地点。走在医院大厅里的访客需要通过标识来获得安心，保证自己走在正确的方向上。如果无法满足客户的预期，他们会在陌生的环境中迷路或感到缺乏确定性。

语言寻路是拼图的另一部分。对需要进一步导航指示的用户或是不懂标识上语言的用户来说，能提供口头帮助的信息台十分重要。前台服务人员应当能以彬彬有礼、始终如一的客观方式呈现信息，减少用户的困惑。视觉寻路元素对约90%的用户都是有效的，而语言寻路元素则主要帮助剩余的10%。在人流密集的医院，它们每年能为成千上万的患者和访客提供帮助。

虚拟寻路设计包含动态、非静止的导航工具；从本质上说，就是数字工具：具有方向信息的电子显示屏、带有外语辅助的交互指南、带有实时停车位计数的"智能车库"标识、智能手机的导航应用程序以及各种其他技术。虚拟寻路元素不断进步，它们将变成未来综合导视系统中至关重要的组成元素。

3V元素提供并强化了同样的信息；由于人们处理信息的方式不同，这些信息以不同方式呈现并产生作用。当视觉标识不充分的时候，语言寻路元素可以填补空白；当虚拟信息不充足的时候，建筑标识可以提供引导。只要各个元素的相互交流，保证相互之间没有矛盾或遗漏，各种寻路元素就能呈现出一致的信息。打造一个能让3V元素相互协作的整体规划十分重要。该策略能保证客户将信息传达到更多的用户，使他们快速、简单地找到自己的目的地并理解品牌。针对信息传递的团队协作、创意灵感和最佳实践的分析，让我们得以改进设计流程，帮客户实现更好的结果。

第四个V元素是价值（value）。本书中的每个项目背后都有一个清楚认识到交流信息的重要价值的客户。建筑师、室内设计师、土地规划师等设计人员对环境图形设计专业价值的正确认识也十分重要。最终，这种价值实现了用户的高满意度，这是每个企业的追求，也是每个设计的最终衡量标准。

环境图形设计在人们的空间体验中扮演了重要的角色；因此，我们接触的每个项目都能以某种方式影响人们的生活质量。本书主要聚焦于医疗业，他们的用户往往都不太舒适。迷路或迷惑只能增加他们的负担。本书中的每个设计都不仅服务于客户，还服务于他们的用户，力求用创新方案来解决实际问题。能够看到我们的设计是如何一天天改变世界，真是我们工作的最大收获。能够被收录在本书中是对我们工作的极大肯定，让我再次祝贺本书中各个作品的设计者们！

吉姆·哈丁
美国，GS&P设计公司

CONTENTS
目录

002 Preface 前言
006 Children's Hospital 儿童医院
068 General Hospital 综合医院
106 Specialised Hospital 专科医院
134 Medical Centre 医疗中心
186 Clinic 诊所
220 Pharmacy 药房
238 Index 索引

Children's Healthcare of Atlanta – Feature Walls at Egleston Campus

Design agency: Stanley Beaman & Sears
Photography: Jim Roof
Client: Children's Healthcare of Atlanta
Country: USA

The project is a part of a $344 million master plan expansion and renovation of both hospital campuses. Project components include a new main lobby, public waiting areas, acute and ambulatory clinical areas, diagnostic and treatment areas, and patient / family support spaces. Graphics, signage, and these interactive features were closely integrated to the comprehensive interior design and concepts. These feature walls incorporate a selection of bold colours, fun textures and finish materials which effectively harmonise with the Interior Design. These interactive walls are inherent to the global success of the project. The solution is one of patient focused care tailored to children and people of all ages that is spirited, welcoming, and educational.

亚特兰大儿童医院——埃格尔斯顿院区特色墙

项目是耗资3.44亿美元的多院区综合扩建和翻修工程的一部分。项目包含一个新大厅、公共候诊区、急诊和门诊区、诊断和治疗区和患者/家属辅助空间。图形、表示和交互特征与整体室内设计和概念紧密结合。这些特色墙的设计选择了大胆的色彩、有趣的纹理和装饰材料，与室内设计和谐统一。这些交互墙壁对整个项目的成功至关重要。设计以患者为中心，为儿童量身定制，各个年龄段的人群都能从中获得启发并感受到热情。

设计机构：SBS设计公司 摄影：吉姆•鲁夫 委托方：亚特兰大儿童医院 国家：美国

Children's Hospital

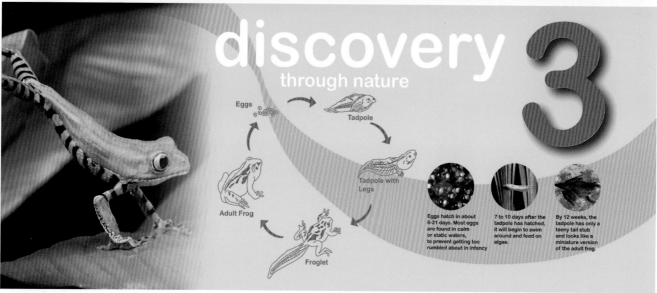

Children's Hospital

Children's Healthcare of Atlanta – Stair Gym at Egleston Campus

Design agency: Stanley Beaman & Sears
Photography: Jim Roof
Client: Children's Healthcare of Atlanta
Country: USA

>>

The client had expressed their desire to communicate their commitment to wellness, not just to their patients, but also to the staff that care for them. In association with one of the organisations very motivated internal health and wellness programs, named Strong 4 Life, the client hired the architectural firm to provide design concepts and then implement the facilities first Stair Gym. The team was challenged with a tight budget to design and implement the entire stair renovation. This would need to include all graphics, signage, paint and flooring. The team met with the client to prioritise the essential items that would make this stair successful. Needing to cut back on audio, sound, lighting and other sensory concept supplements that was presented, the team focused their efforts on the basics of colour, graphics and signage. Floor identifiers were a nicely subtle design. Constructed out of transparent plastic, these budget conscience signage were mounted directly in front of the larger wall graphics.

亚特兰大儿童医院——埃格尔斯顿院区楼梯设计

客户希望传递出自己致力于健康事业的理念。客户与积极上进的健康项目组织"强健生活"（Strong 4 Life），委托建筑事务所为其提供设计概念并将其应用在项目设施中。设计团队必须以有限的预算来完成整个楼梯间的翻修改造，其中包括图形、标识、喷装和地面整修。设计团队与客户确认了重点元素，以确保成功完成项目。设计团队削减了声光等感官概念，将重点放在色彩、图形和标识上。楼层标识符采用了精致的设计。由透明塑料制成的标识物美价廉，被直接安装在大幅墙面图形上。

设计机构：SBS设计公司　摄影：吉姆·鲁夫　委托方：亚特兰大儿童医院　国家：美国

Did you know that by climbing all 5 floors of the Stairgym 4 times per week, it is almost the same as climbing Mt. Everest?

Children's Hospital

Children's Hospital

Children's Healthcare of Atlanta – Donor Recognition Scottish Rite

Design agency: Stanley Beaman & Sears
Photography: Jim Roof
Client: Children's Healthcare of Atlanta
Country: USA

>>

These interactive feature walls are a part of the donor recognition program for a leading national Children's Hospital in Atlanta. Ranked the 3rd best Children's Hospital in the nation and strictly for non-profit, this metro Atlanta hospital resides on one of two large campuses totaling over 1million square feet. Expressing a strong vision of excellence for their project that would be a reflection of their standard of healthcare, the client desired a non-traditional donor recognition feature that would have high visual impact and also be entertaining, educational and appealing. Intentionally nestled at the ends of public corridors and high traffic public areas, these electronic and interactive feature walls serve as both donor recognition and child entertainment interests. Their dynamic characteristics and educational content, successfully respond to the clients desire to engage their young patient population and accompanying families in a playful and non oppressive manner.

亚特兰大儿童医院——苏格兰共济会感恩墙

这些互动特色墙是亚特兰大儿童医院的感恩项目的一部分。作为美国排名第三的非营利儿童医院，该院共有两个院区，总面积超过9.3万平方米。客户要求项目的设计可以反映他们高超的医疗水准，希望打造一个非传统型感恩墙系统，使其兼具视觉影响力、娱乐性、教育性和吸引力。这些电子互动背景墙被有意设置在公共走廊的两端和人流密集的公共区域，既对捐赠者表示了感谢，又能吸引儿童。他们的动感特色和教育内容成功地满足了客户想要吸引小患者的期望，让患者家庭置身于一个没有压力、只有乐趣的环境。

设计机构：SBS设计公司 摄影：吉姆·鲁夫 委托方：亚特兰大儿童医院 国家：美国

Children's Hospital

Children's Hospital

Joe Dimaggio Children's Hospital

Design agency: Stanley Beaman & Sears
Photography: New York Focus, LLC.
Client: Memorial Healthcare System
Country: USA

Believing that the hospital should "look alive", feel non-institutional and inspire hope, "The Power of Play" was chosen as the conceptual story for the building. The story is told throughout the hospital's four floors, each of which focuses upon a specific type of play: the first floor being sports, the second – arts, the third is games, and the fourth is dreams. Each level incorporates a unique primary colour, features a related graphic pattern, and showcases a custom designed silhouette of a child engaged in the type of play for that floor. The silhouettes are utilised in the building's wayfinding as memorable symbols for each floor. On the medical-surgical floor, where it is generally desirable to encourage patients to get up and get moving, the motif is "games". The colour palette of primarily high-energy orange and patterns of squares referencing game boards are integrated into floors and wall graphics. More than simply bright and energising, they are used by staff as tools to encourage their young patients to engage in a therapeutic walk or activity. On another floor, within the oncology unit where patients can tend to be less active and yet still require great hope and encouragement, the play is "dreams". Here, the palette is softer blues, the pattern is of waves, and the silhouette is of a child in a cape dreaming of being a superhero who can fly!

狄马乔儿童医院

设计师坚信医院应该"看起来鲜活"、不过分呆板、能燃起人们的希望,因此选择了"玩的力量"作为建筑的主题。该主题贯穿了医院的四层楼,每层楼聚焦于一种玩法:一楼是体育运动,二楼是艺术,三楼是游戏,四楼是梦想。每层楼都有其独特的色彩,配有相关图画,以特定的儿童剪影为标志。这些剪影被应用在建筑的导视系统中,作为各个楼层的记忆性象征。在医学外科楼层,医生通常鼓励患者站起来运动,因此以"游戏"为主题。以高能的橙色为主的色调和棋盘的方格图案被融入了楼面和墙壁图形的设计中。它们不仅亮丽而充满活力,还是医护人员鼓励小患者参与治疗性散步及活动的工具。在肿瘤科所在的楼层,患者可能不太活跃,但是仍需要大量的希望和鼓励,因此以"梦想"为主题。这里的色彩以柔和的蓝色为主,儿童剪影是一个正梦想成为会飞的超级英雄小孩。

设计机构:SBS 设计公司 摄影:纽约焦点公司 委托方:纪念医疗系统 国家:美国

Children's Hospital

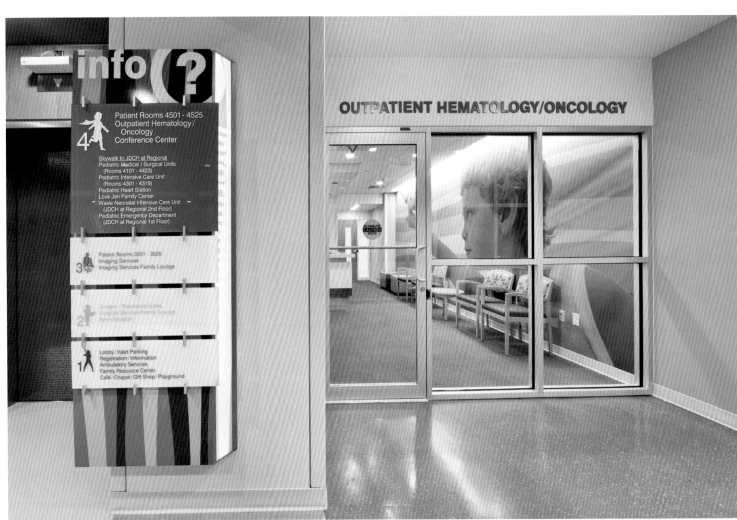

Children's Hospital

Children's Hospital

Nemours Children's Hospital

Design agency: Stanley Beaman & Sears
Photography: Jonathan Hillyer
Client: Nemours Foundation
Country: USA

>>

Wayfinding and feature graphics for the new, 60-acre pediatric health campus, including site, hospital, clinic and parking deck.

Objectives:
Nemours Children's Hospital, recognising the healing value of nature and family involvement, deserved a wayfinding system that:
- reduces stress via its simplicity and ease of use
- promotes accessibility and encourages families' presence
- reinforces the "hospital in a garden" concept with playful, nature imagery that also provides wayfinding cues

Special Features:
Special architectural features throughout the project were used to provide strong visual cues via form, colour, and imagery. For example, the folded planes which draw visitors into the elevator areas are wrapped with large graphic images which fade seamlessly into the wall colour, identifying the floor and integrating both graphic and architecture.

内穆尔儿童医院

项目为新建的占地约24公顷的儿科院区（包含场地、医院、门诊和停车场）提供了导航和特色图形设计。

目标：
内穆尔儿童医院认可自然与家庭对治疗的价值，需要这样的导视系统：
- 能通过简单易懂的应用减少患者的压力
- 提升医院的通达性，鼓励家属出入
- 通过趣味自然图形强化"花园医院"的概念，同时提供导航信息

设计特色：
项目利用建筑特色从造型、色彩和图像三方面提供了强烈的视觉线索。例如，引领访客进入电梯区的折叠平面被大型图片所覆盖，与墙壁的色彩自然融合，标识出楼层，同时也整合了图形与建筑。

设计机构：SBS设计公司 摄影：乔纳森·希利尔
委托方：内穆尔基金会 国家：美国

Children's Hospital

Patient Rooms
- ↑ 5101 - 5108
- → 5109 - 5116
- → 5117 - 5124

Patient Rooms
- ↑ 3101 - 3116
- → 3117 - 3125

Children's Hospital

Palmetto Health Children's Hospital

Design agency: Stanley Beaman & Sears
Photography: Jim Roof
Client: Palmetto Health
Country: USA

Palmetto Children's Hospital in Columbia, South Carolina was the complete interior renovation of an existing adult cancer facility. The wayfinding and graphics in conjunction with the interior design set a tone of high-quality child focused care. The design team had a strong desire to serve the hospital's young patients and their families by creating a vibrant, engaging, and educationally distracting environment-an environment in which they could clearly navigate around the physical environment easily, and at the same time become distracted or absorbed into something of interest that might capture them emotionally and mentally. The design team based their concepts on research that have proven the use of nature themed images in clinical spaces to reduce anxiety and promote healing. Throughout six floors of the building, the design team developed each floor to incorporate large format digital imagery, educational information, framed artwork and sculpture of local artists that highlight a particular biome of the planet Earth. The floors were assigned themes as follows: First Floor-Aquatic; Second Floor- Rainforest; Third Floor-Grasslands; Fourth floor- Polar; Fifth Floor-Temperate; and Sixth Floor- Desert.

帕尔梅托儿童医院

美国南加州哥伦比亚市的帕尔梅托儿童医院经由一座成人癌症治疗机构改造而成。室内的导视设计为高品质儿科医疗奠定了基调。设计团队强烈希望通过打造一个活跃、美观而具有教育意义的环境来服务小患者和他们的家属，让他们在医院里得到清晰的导航，同时获得情绪上的放松。研究证明，在医疗空间使用自然主题图像能显著减少焦虑、提升治疗效果，设计概念就以此为基础。针对建筑的六层楼，设计团队为每层楼开发了不同的大尺寸数码图像、教育信息、带框艺术品和雕塑（由本地艺术家提供）。设计突出了地球上的特定生物群落，每层楼的主题分别为：一楼——水生环境；二楼——雨林；三楼——草原；四楼——极地；五楼——温带；六楼——沙漠。

设计机构：SBS设计公司 摄影：吉姆·鲁夫 委托方：帕尔梅托医疗集团 国家：美国

Children's Hospital

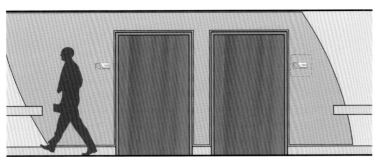

Location Elevation View
Scale: 3/8"= 1'-0"

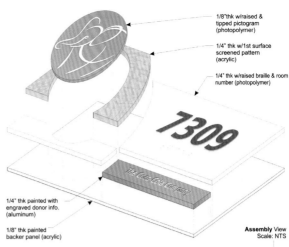

1/8"thk w/raised & tipped pictogram (photopolymer)

1/4" thk w/1st surface screened pattern (acrylic)

1/4" thk w/raised braille & room number (photopolymer)

1/4" thk painted with engraved donor info. (aluminum)

1/8" thk painted backer panel (acrylic)

Assembly View
Scale: NTS

Patient room plaques incorporate unique icons based upon the floor on which they occur: 1st Floor - Aquatic, 2nd Floor - Rainforest, 3rd Floor - Grasslands, 4th Floor - Polar, 5th Floor - Temperate, 6th Floor - Desert.

NOTE: Icon/Pattern for each sign is noted in the **Symbol** column of the **Message Schedule**. Color scheme is per the table noted below and on the following sheet. Pattern color scheme consists of a background of the floor color noted with a foreground pattern of 60% of the noted floor color. Fabricator to submit samples of all finishes for approval prior to manufacture.

Zebra Patient Wing

Giraffe Patient Wing

Cheetah Patient Wing

Scale: 3"= 1'-0"

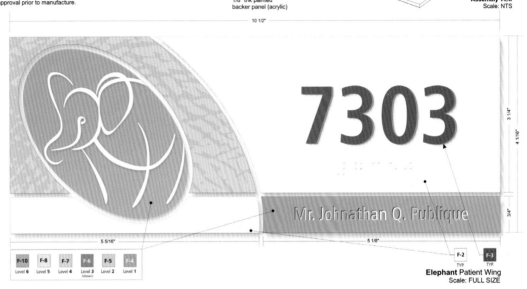

Elephant Patient Wing
Scale: FULL SIZE

F-11 Desert Burnt
SW6883 Raucous Orange LRV 18%
(ARCH n/a)

F-8 Temperate Mint
Duron 7202W Lifting Green
(ARCH P-16)

F-7 Polar Violet
SW6821 Potentially Purple LRV 63%
(ARCH P-13)

F-6 Grassland Blue
SW6962 Dazzle LRV 23%
(ARCH P-10)

F-5 Rainforest LT Green
Duron 7204D Spring Chartreuse
(ARCH P-6)

F-4 Aquatic Blue
SW6486 Reflecting Pool LRV 39%
(ARCH P-4)

GILA MONSTER

DEER

POLAR BEAR

ELEPHANT

TREE FROG

JELLYFISH

ARMADILLO

RACCOON

PENGUIN

ZEBRA

MONKEY

DOLPHIN

DESERT TORTOISE

BALD EAGLE

ARCTIC FOX

CHEETAH

TOUCAN

SEA TURTLE

DESERT CAMEL

PORCUPINE

SEAL

GIRAFFE

TIGER

SEAHORSE

Children's Hospital

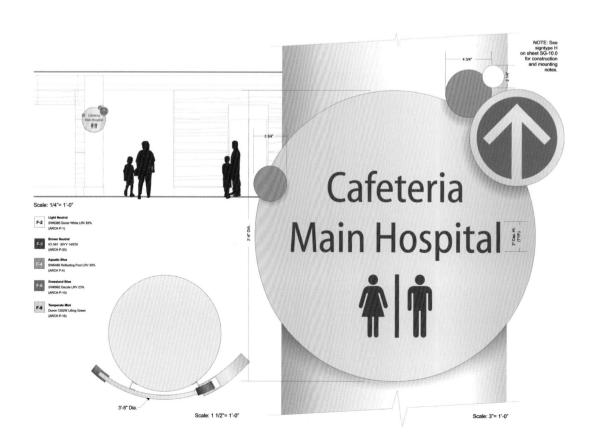

Children's Hospital

Great Ormond Street (Children's) Hospital, London, UK

Design agency: SMITH UK LTD
Designer: Paul Oldman
Client: Great Ormond Street (Children's) Hospital, London, UK
Country: UK

>>

The objective of this project was to deliver more children and family friendly surroundings to a series of new wards that were being added to this world-renowned children's hospital in London. Patients and visitors are often under pressure and often distracted and confused in hospital environments. Navigating your way round a large and often new and unknown environment quickly, easily and without recourse to asking for help can only help alleviate stress. Naming wards and finding interesting and engaging ways to point people in the right direction is a process known as wayfinding. A strategy for wayfinding at GOSH based on the natural world was established where each floor took on a different habitat inhabited with different animals – creating a clear and easy to understand non-language based visual distinction between floors and wards. Smith developed a set of animal characters providing the graphics for ward signs and wall murals throughout the new wards. Smith also introduced new characters and other illustrative elements to enhance the therapeutic environment and ultimately aid recovery. In total Smith delivered graphics for six ward environments for Squirrel, Walrus, Bear, Koala, Flamingo and Eagle.

英国伦敦大奥蒙德街儿童医院

项目的目标是为这座国际闻名的儿童医院新增的病区打造适合儿童及家属的环境。在医院环境中，患者和访客总是压力很大、心烦意乱。如果能在一个全新而未知的环境中提供快速清晰、无需询问他人的导航系统，可以显著减少他们的压力。为病区起名并利用有趣的方式引导方向则称之为导视设计。在大奥蒙德医院的导视设计中，每个楼层被分配作为不同的物种栖息地，形成了简单易懂的视觉区分，无需语言基础。Smith公司开发了一系列动物形象，作为病房标识和墙壁装饰的图形元素。他们还引入了新角色和其他插画元素来改善治疗环境，从而帮助患者恢复。Smith公司为六层病房楼分别开发了六种动物形象：松鼠、海象、熊、考拉、火烈鸟和鹰。

设计机构：SMITH英国公司 设计师：保罗·奥德曼 委托方：英国伦敦大奥蒙德街儿童医院 国家：英国

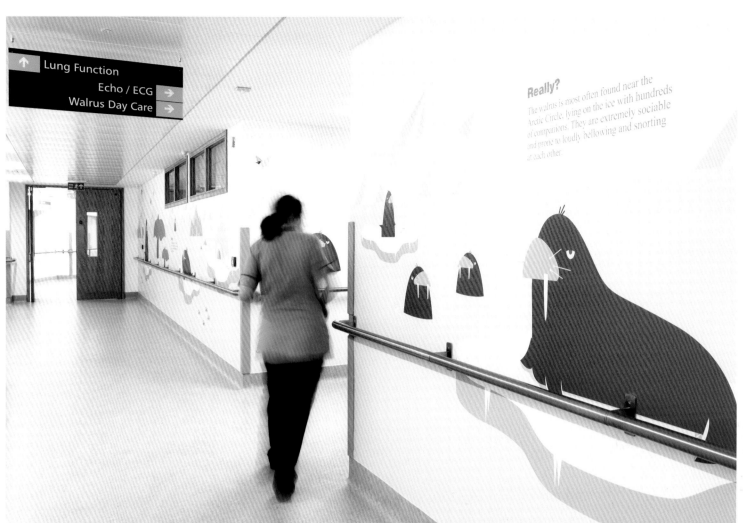

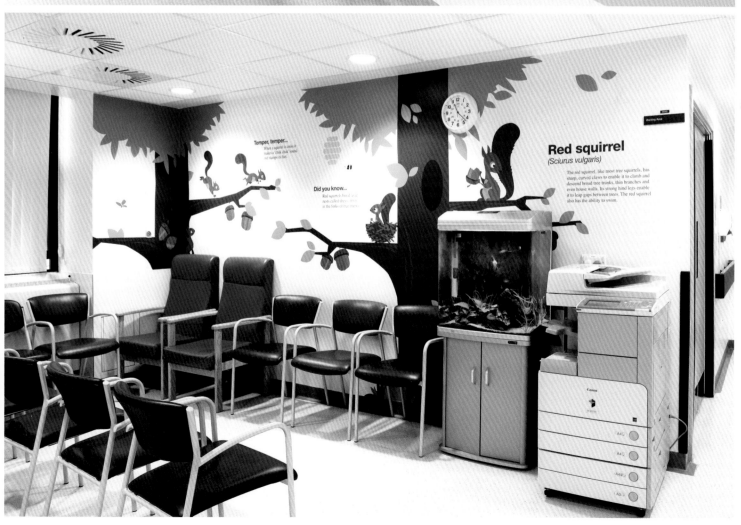

Children's Hospital

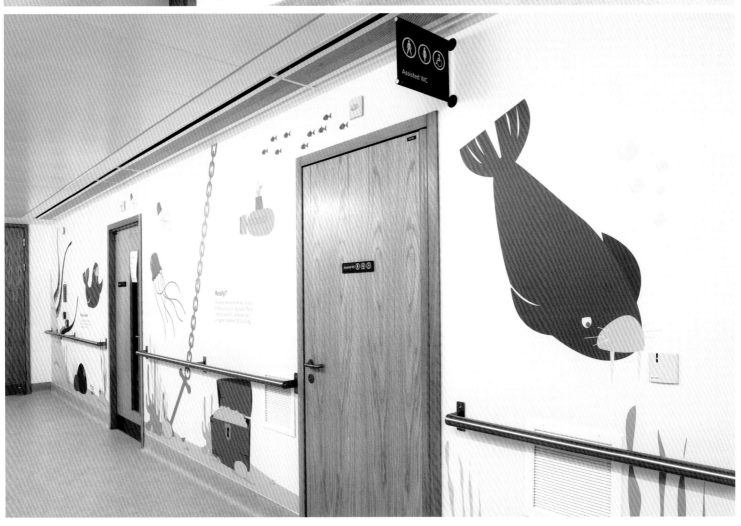

Children's Hospital

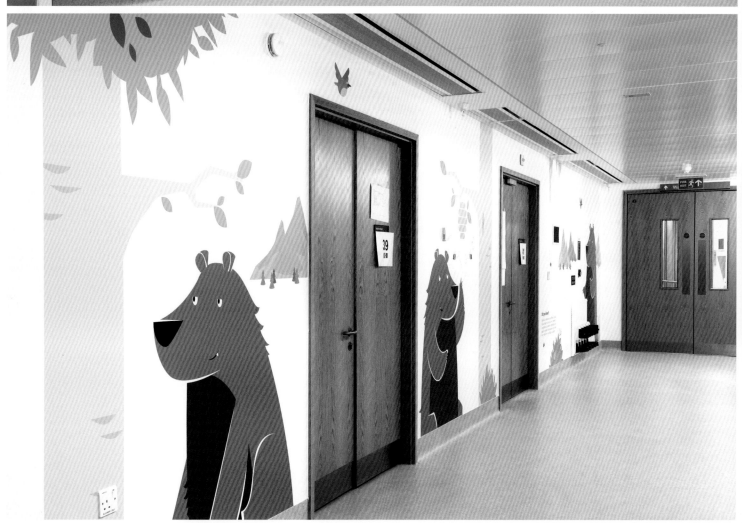

Children's Hospital

Surya Child Care

Design agency: Leaf Design Pvt. Ltd.
Designer: Vijay Poojari
Client: Surya Child Care
Country: India

Surya Child Care is a specialised hospital that has set a benchmark in Neonatal Intensive Care and Pediatric Health Care. As their brand custodians, they approached Leaf Design to round up the branding with holistic wayfinding. The designers used multiple visual media like colours, contours and pictograms to create that perfect environment and crafted an experience which has communication for the big and little, infusing loving, homely and childlike atmosphere which interacts with the kid but gains the trust of parents. The design engages the kids through simple puzzle type game with fun and happy visuals, making it a place where kids love to be. The guide is well informed and reduces the level of anxiety. Signage is very simple, clean, soft and functional. As said small touches of colour makes it more colourful than having the whole thing in colour. So a sandwich structure with respective floor colour formed the language for signage. Pictograms were essential part of the wayfinding system as there was no secondary language in use. Pictograms are designed to be self explanatory, relevant and follow a same tone across.

苏里亚儿童医院

苏里亚儿童医院是一家专科医院,它树立了新生儿重症监护和儿科医疗的标杆。作为它们的品牌管理人,Leaf 设计公司受委托为医院打造了全套的导视系统。设计师利用色彩、轮廓、图标等多重视觉媒介来营造完美的环境。设计为成人和儿童提供了简单的交流方式,让家一般温馨的环境与孩子互动并获得家长的信任。设计通过色彩鲜艳的简单字谜游戏使孩子们融入其中,打造他们所喜爱的空间。导航信息完整良好,能够减少患者及家属的焦虑感。引导标识简单、清晰、柔和、实用。小面积的色彩点缀比整体实用彩色更加显眼,每个楼层被赋予了独立的楼层色彩。图标是导视系统的主要构成部分,因为项目并没有采用第二种语言。图标的设计简明易懂,所有图标全部采用了统一的基调。

设计机构:Leaf 设计公司 设计师:维贾伊·普加里 委托方:苏里亚儿童医院 国家:印度

Labour Delivery Suite

C4

Dr. Vishal Baldua
MD (Paed), Fellowship in Intensive care (Canada)

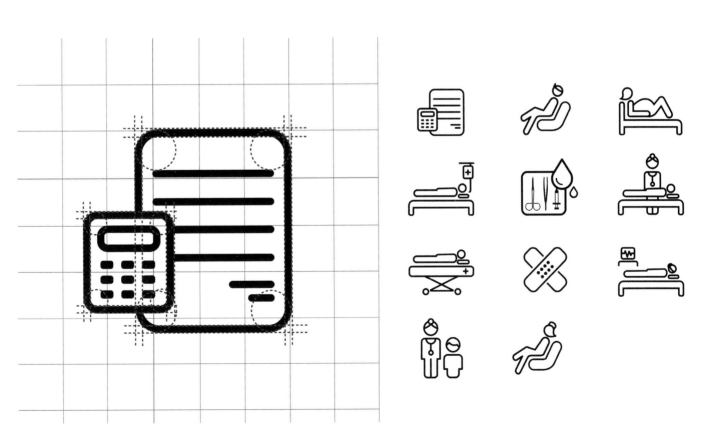

Children's Hospital

The Royal London Children's Hospital

Designer: Morag Myerscough, Donna Wilson, Miller Goodman, Tord Boontje, Chris Haughton, Ella Doran
Client: Commissioned by Vital Arts, Barts Health NHS Trust.
Country: UK

The Royal London Children's Hospital officially opened in March 2012 and over the past two years Vital Arts have been working with a variety of critically acclaimed artists and designers to transform the walls of the children's wards through pioneering and innovative design within a hospital environment. With 130 beds across five wards and covering London's busiest paediatric Accident & Emergency department, The Royal London Children's Hospital, run by Barts Health NHS Trust, is one of the leading children's hospitals in the UK, caring for more than 40,000 children and young people each year from London, Essex and across the UK and Europe. It is well documented in medical research that children find visiting hospital, no matter how routine the visit or how often they come, a frightening and stressful experience. Art has proven throughout paediatric hospitals all over the world, to be an invaluable aid for staff in either distracting children, or positively engaging children receiving sometimes painful and frightening treatments by supporting the individual needs of each patient. Through consultation with medical staff and patients Vital Arts have commissioned artists and designers that showcase dynamic design whilst meeting the patient's needs.

皇家伦敦儿童医院

皇家伦敦儿童医院于2012年3月正式开放，在过去两年多的时间里，Vital Arts基金会与各路广受好评的艺术家和设计师将儿童住院区的墙壁通过先锋创新设计进行了改造。由巴兹保健和国民信托运营的皇家伦敦儿童医院拥有130张病床，拥有伦敦最繁忙的儿科急诊部，是英国最顶尖的儿童医院之一，每年从伦敦、埃塞克斯郡以及英国和欧洲各地收治的患儿可达40,000多人次。医疗研究表明，即使是经常造访医院的儿童，在医院中仍会感到害怕和紧张。全球各地儿科医院里的艺术品为帮助医护人员分散儿童注意力或使儿童积极参与到治疗中能起到辅助作用。在咨询了医护人员和患者之后，Vital Arts基金会委托了艺术家和设计师在满足患者需求的前提下展示了丰富的设计作品。

设计师：莫拉格·麦耶斯考、唐娜·威尔森、米勒·古德曼、托德·布欧尔、克里斯·霍顿、埃拉·多兰
委托方：Vital Arts基金会、巴兹保健和国民信托
国家：英国

Children's Hospital

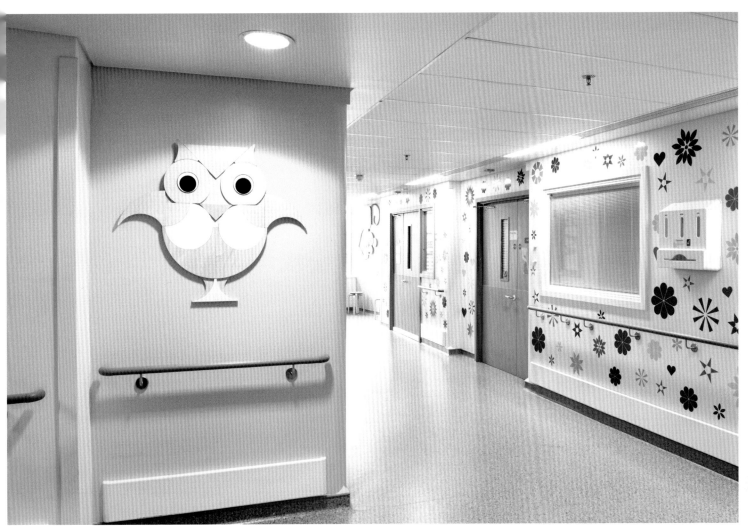

Children's Hospital

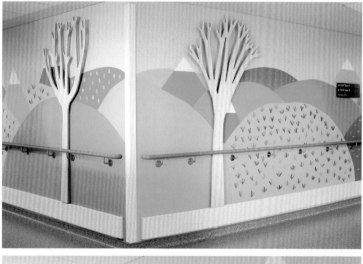

Children's Hospital

Children's Hospital

Protestant Hospital Queen Elizabeth

Design agency: Dan Pearlman
Photography: diephotodesigner.de for :: Dan Pearlman
Client: Evangelisches Krankenhaus Königin Elisabeth Herzberge gemeinnützige GmbH (KEH), (Protestant Hospital Queen Elizabeth)
Country: Germany

The 'Evangelisches Konigin Elisabeth Krankenhaus' Hospital, in Berlin, created a holistic dreamland for young patients with psychiatric issues in cooperation with Dan Pearlman creative agency. The patients of the Department of Child and Adolescent Psychiatry of the 'Evangelisches Konigin Elisabeth Krankenhaus' Hospital (KEH) in Berlin will never feel alone again. In addition to doctors and nurses the youngsters are hosted by Princess Elise. She is the protagonist of the story told to the children when they arrive at the institution: when she was young, Princess Elise invented her own island, with golden sand, palm trees, rocks that kiss the waves and shelters where she could rest when she needed to. Here she was always safe. Now as a grown-up lady, Elise does no longer need to live on the island and therefore dedicated it to her young patients in the hospital. Each room in the hospital was designed in terms of shape, colour, materials and light with the aim of promoting a positive atmosphere, empowering children's imagination and positive emotions. The new concept meets the demands of young patients and therapeutical staff at the same time. The obligations of therapeutic professionals and the emotional needs of young patients are now balanced.

伊丽莎白女王新教医院

在 Dan Pearlman 设计公司的协助下，柏林伊丽莎白女王医院为患有精神疾病的小患者打造了一个梦幻空间。儿童与青少年精神疾病部的患者永远都不会感到孤单。除了医生和护士之外，患者还被"艾莉丝公主"所关心。当患儿到达医院后，医院会告诉他们艾莉丝公主的故事：小时候，艾莉丝公主建立了自己的岛屿，岛上有金沙、橄榄树、岩石、海浪和小屋。她在这里是安全的。现在，作为一名成熟女性，艾莉丝不再需要住在岛屿上，因此她将它奉献给了医院的患儿。医院的每个房间在造型、色彩、材料和灯光等设计方面都力求营造一种积极的环境，让儿童充分想象，释放积极的情绪。新的设计概念满足了患儿与医护人员的双重需求。医护人员的责任和患儿的情绪需求得到了良好的平衡。

设计机构：Dan Pearlman 设计公司 摄影：diephotodesigner.de 摄影 委托方：伊丽莎白女王医院 国家：德国

Children's Hospital

Children's Hospital

Jingdu Children's Hospital

Design agency: StudioSigno
Design director: Long Ping, Wu Yan
Illustrator: Zheng Chuyang
Space Designer: Chen Jiantong
Graphic Designer: Cui Tan
Country: China

Jingdu Children's Hospital is one of the biggest 3 children's hospitals which based in Beijing, China. In this project, StudioSigno designed the environmental graphic design of Jingdu Children's hospital, which is also the first approach of applying environmental graphic in China. StudioSigno created three individual animal figures. The theme "realising dream together" runs through the hospital's design. The floors are divided as ocean, desert and prairie to provide children fun experience. Each department is equipped with a lovely icon, attractive to children. The design not only remains on flat surface, but also runs through the whole space: columns are turned into coral trees; wards are added with plots; runways are set on walls and ground to invite children to participate "animal sports games"... These design elements are also transformed as 3D entities to recreation facilities, toys and furnishings. StudioSigno uses design to inspire children's courage to fight diseases and to reduce their resistant feeling to hospital and treatment.

京都儿童医院

京都儿童医院是北京三大儿童医院之一，汉符设计为其提供了全套的环境图形设计，这也是中国首个环境图形设计项目。汉符创造了三个个性鲜明的动物形象，以三个好朋友共同"实现梦想"为主线，贯穿医院整体设计。将医院空间以楼层区分为海洋、沙漠、草原等不同故事场景，让孩子进入医院便是一种趣味体验。并特别为每个诊疗科室设计了可爱的图标，对儿童充满了吸引力和趣味性。设计并不仅停留于平面，他们充分利用空间本身，将柱子演变成珊瑚树、为病房植入故事情节、在墙面与地面上设置跑道，让孩子一起参与到"动物运动会"中……这些设计还被转化为三维实体延展至儿童游乐设施、玩具、家具等。汉符用设计激发儿童战胜疾病的勇气，减少对医院与治疗的抗拒。

设计机构：汉符设计 设计总监：龙平、吴燕 插画师：郑初阳 空间设计师：陈建同 平面设计师：崔谭 国家：中国

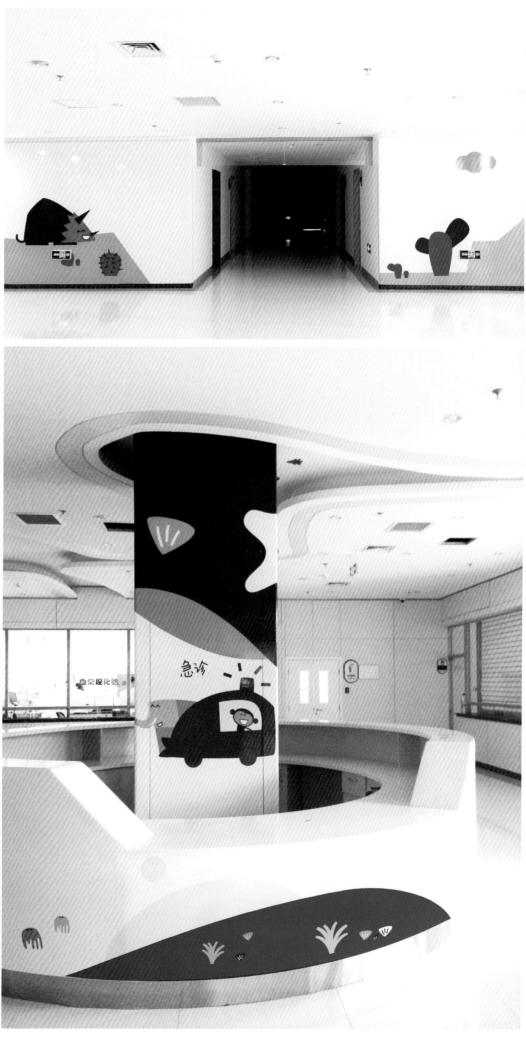

Children's Hospital

Jingdu Children's hospital is one of the biggest 3 children's hospitals which based in Beijing, China.
In this project, we designed the environmental graphic design of Jingdu Children's hospital, which is also the first approach of applying environmental graphic in China.

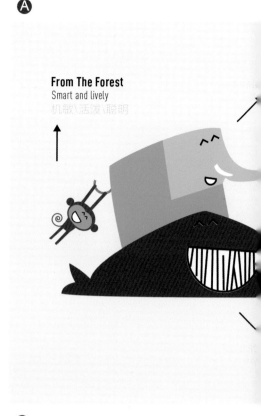

From The Forest
Smart and lively
机敏\活泼\聪明

Ⓐ Main characters
Three main characters – whale, elephant and monkey, which are attrative and intresting for children, decrease their fear of hospital or disease.

Ⓑ Icons of clinic departments
Specially designed lovely icons for each department.

Ⓒ Track
Not only to be friendly, but also to tell children to be brave and to try their best to go for their aims.

Ⓓ Themes & storylines
In a general frame of 3 characters keep their dreams alive, each theme has its own story. For example, the elephant and the monkey help the whale to make his dream true — to go travel on the land.

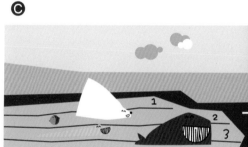

Ⓔ Furnitures
Stories based furnitures were also designed, in order to bring a rich experience to children.

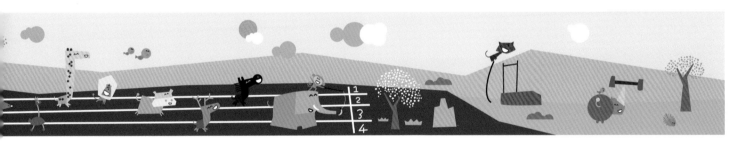
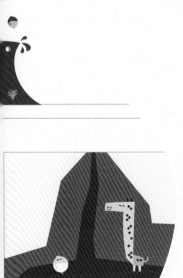 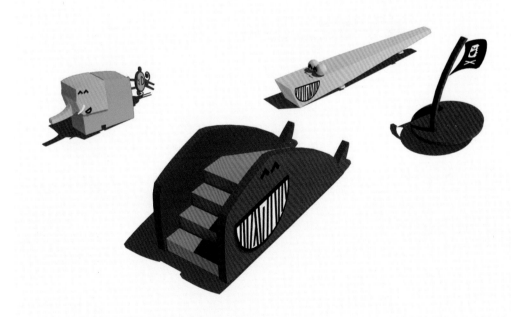

Children's Hospital

Lady Cilento
Children's Hospital

Design agency: Dotdash
Designer: Domenic Nastasi
Photography: Dianna Snape, Christopher Frederick Jones
Client: Conrad Gargett Lyons
Country: Australia

>>

Dotdash have designed and integrated a wayfinding strategy which includes a collection of signage and environmental graphics to connect a network of spaces, and provided a cohesive system of navigation throughout complex external and internal environments. The family of wayfinding components work in unity to distract from the nature of the hospital environment by expressing a bold, vibrant aesthetic. Each component aims to lighten the experience of the facility to its younger visitors.

齐兰托夫人儿童医院

Dotdash 设计公司为医院设计并整合了导视策略，包括一系列标识与环境图形。他们还为复杂的室内外环境提供了一套综合的导航系统。导视设计元素通过大胆、活泼的美学转移了人们的注意力，弱化了医院环境的本质。每个元素的设计目标都是减轻小患者的焦虑感。

设计机构：Dotdash 设计公司 设计师：多梅尼克·纳斯塔西 摄影：戴安娜·斯内普、克里斯多夫·弗雷德里克·琼斯 委托方：康拉德·加吉特·莱昂斯 国家：澳大利亚

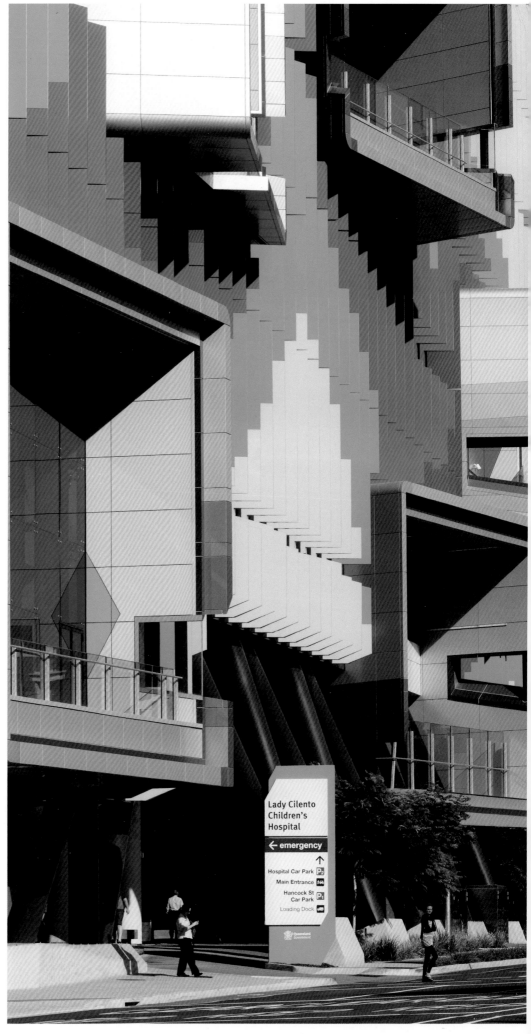

Children's Hospital

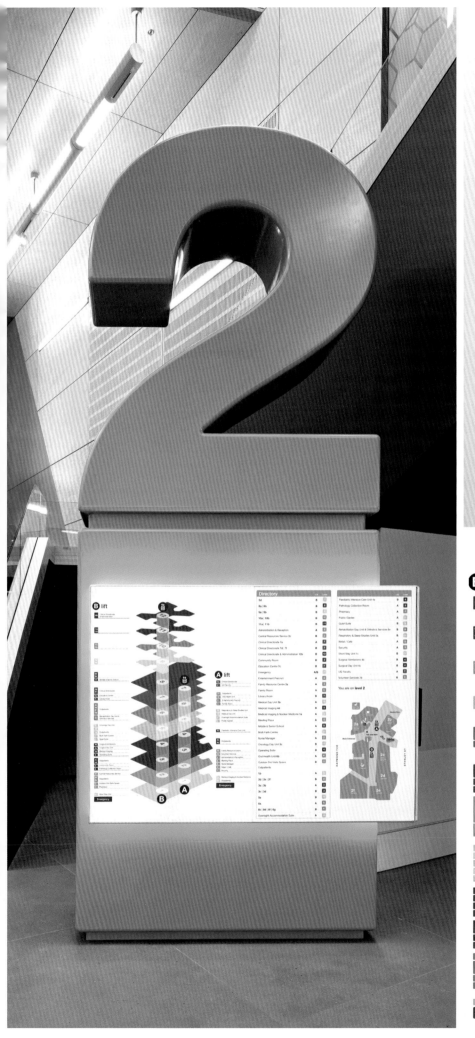

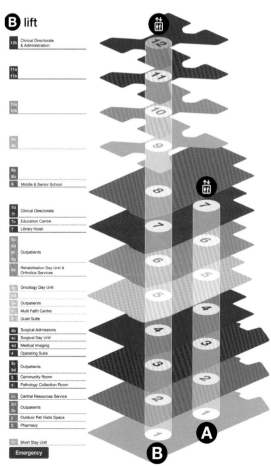

Children's Hospital

Children's Hospital

Hospital Mater Dei

Design agency: Greco Design
Designer: Gustavo Greco, Rafael Sola
Client: Hospital Mater Dei
Country: Brasil

Wayfinding system for Mater Dei Hospital. In dealing with a large-sized building, which may undergo future alterations in its internal flow, a modular signage system makes the exchange of information easier, while furnishing integrity to the system as a whole. Mater Dei's symbol, a cross made with injection molded ABS plastic, is the main linking point between all the equipment pieces. The pictograms, whose designs follow the elements of Myriad – the logo's typography, deliver even greater singularity and identity to the project.

圣母医院

本项目是为圣母医院设计的导视系统。在大规模建筑（未来可能会进行内部改造）的处理上，模块化导视系统让信息交换更简单，陈设装饰可与其融为一体。圣母医院的象征——由 ABS 塑料注塑而成的十字架——是所有模块设计的线索元素。图标设计采用了 Myriad 字体，与医院标识的字体相同，为项目提供了独特的辨识度。

设计机构：Greco 设计公司 设计师：古斯塔沃·格雷科、拉斐尔·索拉 委托方：圣母医院 国家：巴西

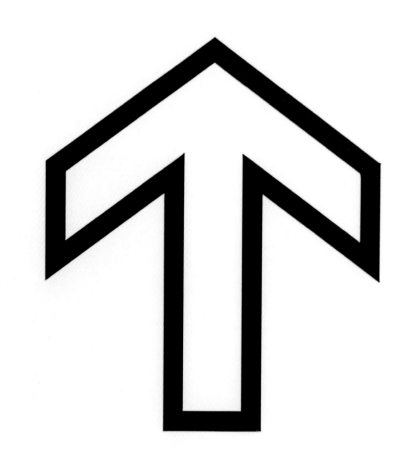

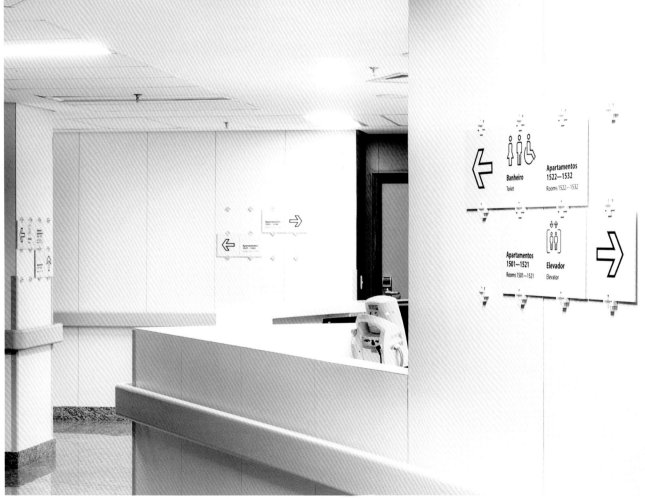

General Hospital

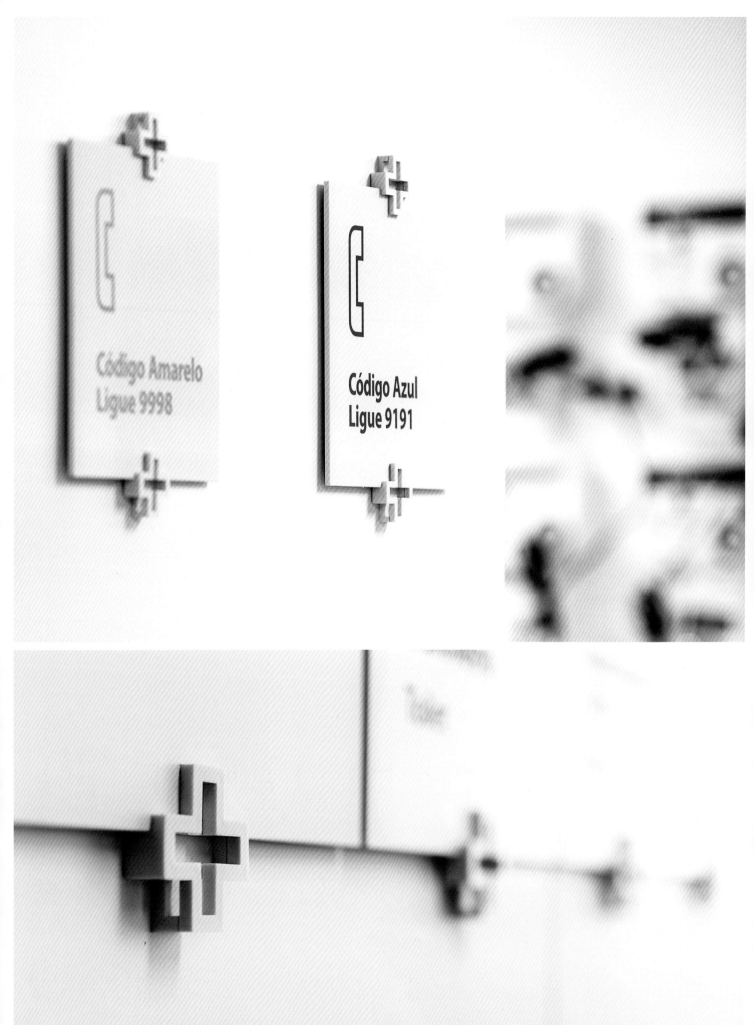

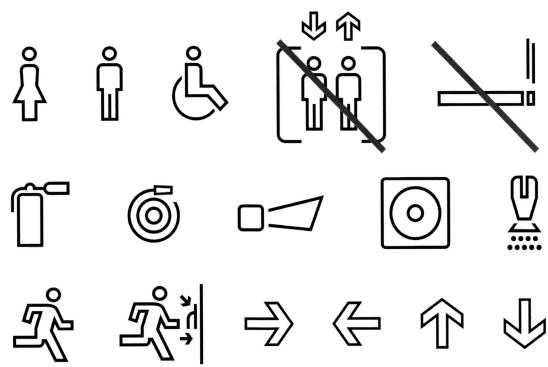

General Hospital

Danbury Hospital: Neonatal Intensive Care Unit (NICU)

Design agency: Perkins Eastman
Photography: Chris Cooper
Client: Danbury Hospital, Western Connecticut Health
Country: USA

The new state-of-the-art NICU – built according to the Planetree model of care – was designed to accommodate not just growth, but also a new quality of care that would far surpass any other in the state. The NICU unit is divided into four pods with the themes earth, fire, water, and air incorporated via glass walls at the entrance to each room, and through signage. Planetree planning concepts include access to information, human interaction, family involvement, nourishment, nurturing environments, and visual therapy. The new space incorporates textures and colours found in nature, creating a series of experiences for parents and staff as they enter the NICU unit with a soft soothing colour palette – helping put the parents at ease. This succession of spaces starts with a large family waiting room and family kitchenette, and the "NICU Graduates Wall" (also called the "Fire Wall" because of its red colour and texture). The wall acknowledges feelings of anxiety parents may experience when entering the unit, and aims to help calm them with pictures of all the NICU graduates. They are then led to the family respite space with a water feature.

丹伯利医院：新生儿重症监护室

这个拥有最先进技术的新生儿重症监护病房根据梧桐树医疗模式（Planetree）建造，将为新生儿带来该州最好的医疗服务。新生儿重症监护病房被分为四个部分，分别以土、水、火和空气为主题，通过入口的玻璃墙和标识进行区分。"梧桐树"规划概念包含信息获取、人际互动、家庭参与、营养、养育环境和视觉疗法。新空间组合了自然界中的纹理和色彩，为家长和医护员工打造了一系列的体验。柔和舒缓的色彩搭配有助于缓解家长的焦虑。这个连续空间从家属等候大厅开始，逐次是家属厨房、"新生儿监护室毕业墙"（因为红色的色彩和纹理，也被称为"火焰墙"）。墙壁上列出了父母可能经历的焦虑感，通过成功存活下来的婴儿名单来缓解他们的焦虑。家属接待处还设有一个水景装饰。

设计机构：Perkins Eastman 设计公司 摄影：克里斯·库珀 委托方：丹伯利医院、西康涅狄格州卫生保健部 国家：美国

General Hospital

Sarasota Memorial Hospital

Design agency: Gresham, Smith & Partners
Designer: Jim Alderman, Betty Crawford, & Mike Summers
Client: Sarasota Memorial Hospital
Country: USA

Gresham, Smith & Partners provided architectural design, interior design and environmental graphic design services for a new Patient Bed Tower expansion project at existing Sarasota Memorial Hospital, in Sarasota, Florida. Creative solutions to emphasise the new building's clean, simple and contemporary exterior were incorporated into the interior signage by the GS&P environmental graphics designer team. The new design colour scheme also reflects the well-known and recognised "Sarasota blue". The ground floor lobby is a very open and intuitive floor plan welcoming visitors to the patient rooms as well as to the Surgery Centre. Individually cut aluminum letters, vertically oriented, easily guide visitors through the lobby to the appropriate area. Environmental graphics and interiors coordinated the use of inspirational images for each floor. This was accomplished through full colour framed feature signage on each floor and wayfinding signs which repeat the inspirational image in a more subtle and soft creative method. The combination positively affects the visitor and patient experience.

萨拉索塔纪念医院

GS&P设计公司为美国加州的萨拉索塔纪念医院新建的病房楼提供了全套的建筑设计、室内设计和环境图形设计服务。GS&P的环境图形设计团队将建筑简洁现代的外观设计带进了室内标识设计中。设计的色彩方案反映了具有高辨识度的"萨拉索塔蓝"。一楼大厅十分宽敞，直观的楼面布局引导着访客走向病房和手术中心。独立切割的铝制字母被安装在垂直表面上，引导着访客从大厅走向他们的目的地。环境图形和室内设计通过各个楼层的引导图像相互协作。每层楼的全彩边框特色标识和引导标识以更精致、柔和的方式重复了楼层图像。二者的结合能够积极地影响访客和病患的体验。

设计机构：GS&P设计公司 设计师：吉姆·奥尔德曼、贝蒂·克劳福德、麦克·萨默斯 委托方：萨拉索塔纪念医院 国家：美国

EMERGENCY

1'-9"

 8"

Lower edge of building fascia

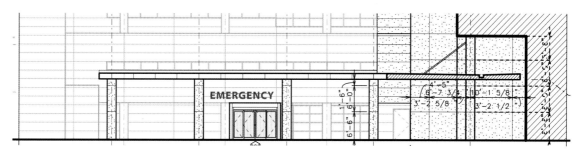

21" letters with raceway, mounted on building fascia, centered visually between canopy columns.

D.2 Internally Illuminated Letters
loc EXT.16

CLEARANCE 13'-6"

1'-0" (field verify) 6" 2"

face of scheduled canopy

← Ⓟ Waldemere Garage | Ⓟ Lasula Garage | Waldemere Medical Plaza ↗

loc EXT.23
Side A (west bound on Waldemere)

EMERGENCY
Hospital Main Entrance

loc EXT.23
Side B (east bound on Waldemere)

General Hospital

Unimed Day Hospital

Design agency: SCENO Environmental Graphic Design
Designer: Gabriel Gallina, Roberto Bastos, Fernando Franco, Douglas Dagostini, Jonas Silveira
Photography: Pedro Milanez
Client: Unimed Vale dos Sinos
Country: Brasil

This is a signage project for a hospital and emergency care where good visual perception of identifying communication, directional and informative was contemplated, where the design features are following the visual identity of the Unimed brand.

巴西医保集团日间医院

本项目是为一所医院及其急诊中心所设计的导视项目。SCENO 为其提供了良好的视觉识别系统以及导向和信息设计。设计特色沿用了巴西医保集团品牌的视觉形象。

设计机构：SCENO 环境图形设计公司 设计师：加布里埃尔·加里纳、罗伯托·巴斯托斯、费尔南多·弗朗科、道格拉斯·达格斯蒂尼、乔纳斯·西尔维拉 摄影：佩德罗·米拉奈斯 委托方：巴西医保集团 国家：巴西

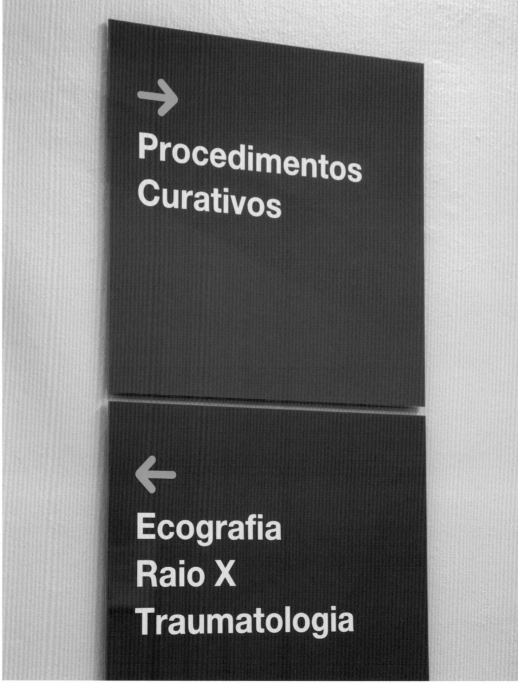

General Hospital

Wockhardt Hospital

Design agency: Leaf Design Pvt. Ltd.
Designer: Vijay Poojari
Client: Wockhardt Hospital
Country: India

Wockhardt Hospital's existence is the result of a 40-year tradition of caring and innovation nurtured by Wockhardt Ltd, India's 5th largest Pharmaceutical and Healthcare company. As Wockhardt is poised to become the most advanced and progressive healthcare institution in India it wanted a branded environment that could walk in step with them into tomorrow. Leaf built upon the timeless belief that good health is a state of mind and tapped into the infinite world of art and expression to create an environment filled with beauty and pleasure. By focusing on admiration which is far from anxiety and suffering, it allowed a few moments of complete relaxation, while the medicine takes care of the body.

沃克哈德医院

沃克哈德医院由拥有40年医疗改革经验的沃克哈德公司所创建，后者是印度第五大制药与医疗保健公司。沃克哈德公司的目标是成为印度最先进、最高级的医疗保健机构，因此，需要一个品牌环境与其共同前往更美好的明天。Leaf设计公司以"健康是一种心态"的理念为基础，运用艺术表现打造了一个充满美感与愉悦感的环境。来访者会将重点放在赞赏而不是焦虑和痛苦上，从而在治疗过程中实现完全的放松。

设计机构：Leaf设计公司 设计师：维贾伊·普加里 委托方：沃克哈德医院 国家：印度

> *Laughter is the sun that drives winter from the human face.*

Our sophisticated technology, advanced medical knowledge and technical and artistic mastery combine to offer an endless array of options that help improve a patient's appearance and function.

There are always flowers for those who want to see them.
—Henri Matisse

Intensive Care Unit 4/5

General Hospital

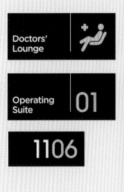

General Hospital

Box Hill Hospital

Design agency: ID/Lab
Photography: ID/Lab
Client: STH Architects
Country: Australia

Box Hill Hospital is a two-staged project. It includes the new building that houses the acute services, while stage two includes the refurbishment of the existing hospital facilities. STH Architects engaged ID/Lab to develop a complete signage package and wayfinding strategy for this major referral centre in Melbourne's East. The strategy needed to connect both buildings and cover the whole of the journey. The font they choose, the amplitude, shows good x-height and has high legibility. The sharp connections reflect the architecture, while one of the environmental graphics was used (tree/leaves treatment) as a basis to develop the signage into softer forms. The result is a clean, approachable striking design.

博士山医院

博士山医院项目分为两期进行施工：一期包括一座新建的急诊楼；二期工程是对原有的医院设施进行的翻修。STH 建筑事务所委托 ID/Lab 开发整套的标识和导视策略。导视策略既要把各个建筑连接起来，又要覆盖整个医院。设计师所挑选的字体 amplitude 具有良好的字高和辨识度。锐利的连接牌反映了建筑特色，而以树木/树叶为主题的环境图形则使其标识设计变得更加柔和。最终的设计简洁、亲切而又显眼。

设计机构：ID/Lab 设计公司　摄影：ID/Lab 设计公司　委托方：STH 建筑事务所　国家：澳大利亚

General Hospital

General Hospital

General Hospital

Cabrini Private Hospital

Design agency: ID/Lab
Photography: ID/Lab
Client: Cabrini Private Hospital
Country: Australia

>>

ID/Lab was initially engaged by Cabrini hospital to conduct a signage audit with a view to making the site more user-friendly. The direct observation from the signage audit was that staff didn't refer to the signs, that there were unrelated sign systems, and there was an inconsistent approach to direction-giving. ID/Lab then developed a new overarching wayfinding strategy and designed signage elements that complemented the intended look and feel of the hospital. The new strategy makes the wayfinding journey more specific and improves legibility. ID/Lab's hardware design solution combines curved panels with high-contrast signage panels. With the addition of timber veneer details, the system steps away from a typical utilitarian hospital feel toward the intended high-end 'hotel' standard of finish.

卡布里尼私立医院

ID/Lab 设计公司受卡布里尼医院委托对医院的导视系统进行了全面审核。在审核中他们发现，员工并不会参考标识，系统缺乏相关性，它们所指引的方向也前后矛盾。ID/Lab 设计公司随后为医院开发了全新的导视策略，设计了一套与医院外观和感觉相匹配的导视元素。新策略让引路设计变得更加详细，改善了可识别性。ID/Lab 设计公司的硬件设计策略结合了弧形板和高对比度标识板。木饰面的加入让导视系统变得与众不同，更具一种酒店装饰的档次。

设计机构：ID/Lab 设计公司　摄影：ID/Lab 设计公司　委托方：卡布里尼私立医院　国家：澳大利亚

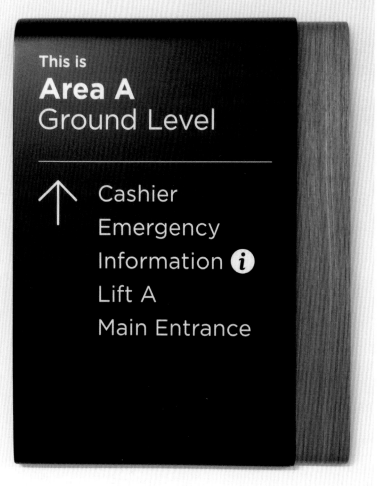

General Hospital

USP Dexeus University Institute, Barcelona

Design agency: clasebcn design
Client: USP Hospitales
Country: Spain

The signage design project combines both, the needs of the users and the concept of the architecture, besides being functional (like any other project in this area). Thus, all the applied elements such as the materials, the colours, the modulations, the dimensions, the proportions and the placement of the signboards respond to the idea to integrate the signage into the hospital in a natural way.

巴塞罗那 USP 迪克瑟斯大学研究院

除了具有实用功能之外，项目的导视设计兼顾了用户需求和建筑概念的表现。引导标识的材料、色彩、调整、尺寸、比例及位置等应用元素全部以一种自然的方式与医院融为一体。

设计机构: clasebcn 设计公司 客户: USP医院 国家: 西班牙

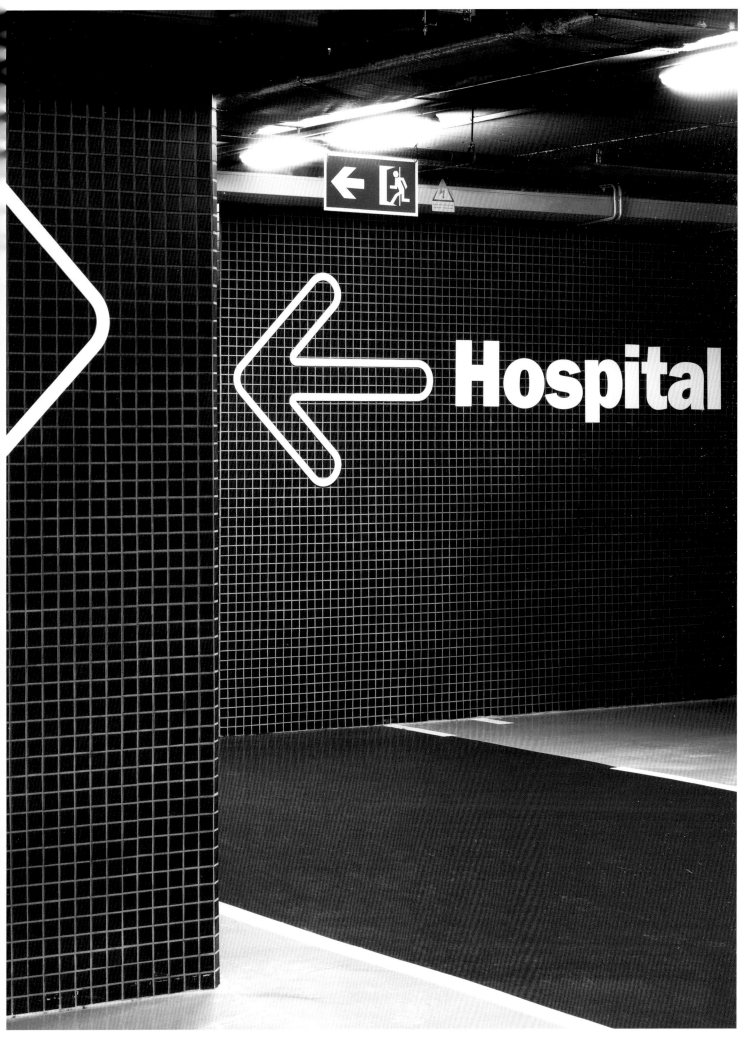

General Hospital

Rocio Hospital

Design agency: MCA Manoel Coelho Arquitetura & Design
Architect: Manoel Coelho, Antonio Abrão, Catia Branco, Andressa Kreusch, Carolina Henares, Andréia Ferrari
Designer: Roberta Perozza
Intern: Fernanda Caroline, Julio Teodoro, Luciana Mayume
Photography: Pedro Coelho
Client: Rocio Hospital
Country: Brasil

This project had patient-focused wayfinding system. The designers wanted clarity of the architectural signage and the wayfinding plan of the place. What's more, the hospital used colour-coding to direct visitors, patients and staff to different. It is an effective wayfinding, and that means straightforward, simple and clear directions.

罗西奥医院

项目是以患者为中心的导视系统。设计师希望建筑标识和引路设计清晰易懂。此外，医院还使用色彩代码来引导访客、患者和员工。导视系统高效实用，直观、简单、清晰。

设计机构：MCA建筑设计公司 建筑师：马诺埃尔·科埃略、安东尼奥·阿布拉奥、卡蒂亚·布兰科、安德萨·克莱西、卡洛琳娜·埃纳雷斯、安德里亚·法拉利 设计师：罗伯塔·佩罗萨 实习：费尔南达·卡洛琳、朱利奥·特奥多罗、卢恰娜·玛由米 摄影：佩德罗·科埃略 委托方：罗西奥医院 国家：巴西

General Hospital

General Hospital

Royal North Shore Hospital Acute Services Building

Design agency: Anne Gordon Design
Creative director: Anne Gordon
Designer: Tim Walker
Photography: Rowan Turner
Client: Thiess
Country: Australia

Wayfinding strategy and signage for the Acute Services Building – Royal North Shore Hospital, Sydney. The recently completed Acute Care Hospital is the largest in Australia. Yellow, green and orange panels highlight the exterior of this enormous building. The vision was to translate those colours into environmental graphics and signage within the building, to compliment the architecture and provide visitors and patients with the wayfinding tools they need in this large hospital. Predominantly colour was applied to each lift core, where lobby ceilings protrude the internal corridor spaces, for recognition and connection to the colour. The directory signage provides the detailed information that visitors use to find their destination, and coloured directional signs reinforce the colour of the lifts and assist with them their journey.

皇家北岸医院急救服务楼

该项目是为悉尼皇家北岸医院的急救服务楼所设计的导视系统。这座新近建成的急救服务楼是澳大利亚最大的同类建筑。黄、绿、橙三色的板材突出了这座宏伟建筑的外观。设计师决定将这些色彩平移到建筑内部的环境图形和导视设计中,既与建筑相匹配,又为患者和访客提供良好的引路工具。主要色彩被应用在各个电梯厅里,大厅的天花板向内部走廊空间伸出,实现了色彩的识别和连接。导航标识为访客提供了详细的信息,而彩色导向标则突出了电梯的颜色,进一步辅助了访客的行程。

设计机构:Anne Gordon 设计公司　创意总监:安妮·戈登　设计师:蒂姆·沃克　摄影:罗恩·特纳　委托方:Thiess 公司　国家:澳大利亚

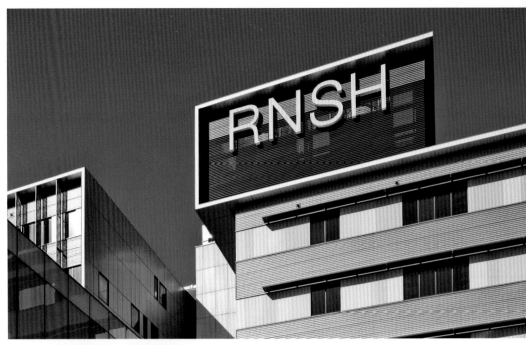

General Hospital

General Hospital

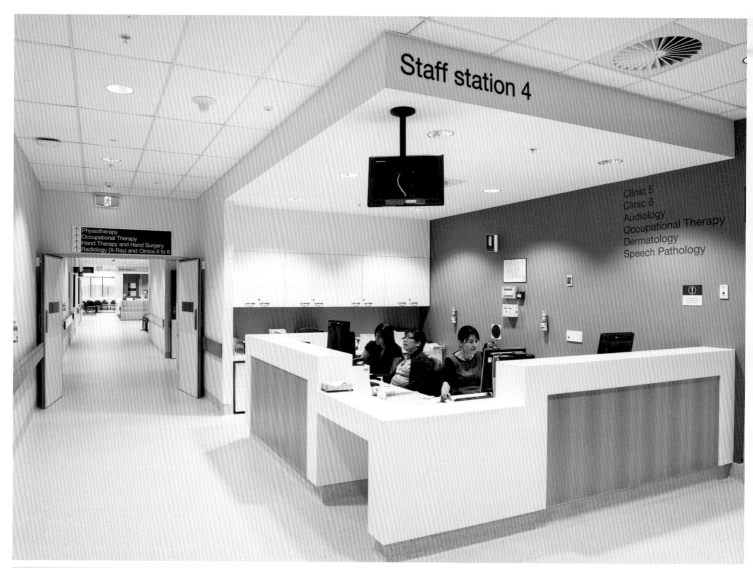

General Hospital

Maternity Hospital Ramón Sarda

Designer: Estefanía Belén Tinto & Victoria Alfieri
Client: Maternity Hospital Ramón Sarda
Country: Argentina

The identity design made to the Maternity Hospital Ramón Sarda was developed to the Rico Design Chair 3, subject that belongs to the Graphic Design course in FADU (UBA). Communication between professionals and patients, as well as their circulation and orientation within the hospital, were taken as problematic issues. In order to accomplish the goals the designers proposed to perform a communication system taking into account the patients' and medical team's characteristics and needs. Environmental graphics and brochures were designed considering not only the context, but also the tastes and interests of each user. As regards the orientation and circulation, a signage system was developed in order to guide the user and to make it understandable. With these resources and tools the designers seek to create a warm and pleasant environment where users feel comfortable and could be recognised as part of the Hospital.

拉蒙萨尔达妇产医院

拉蒙萨尔达妇产医院的识别形象设计属于布宜诺斯艾利斯大学平面设计课程的一个课题。专业人士与患者的交流以及患者在医院中的交通和导向是研究的主要问题。为了实现目标，设计师提议打造一个考虑到患者和医疗团队个性和需求的交流系统。环境图形和宣传册的设计不仅考虑了环境，还考虑了每个用户的品味和兴趣。在导向和交通方面，新开发的导视系统将以简单易懂的方式引导用户。设计师利用这些资源和工具成功地营造了一个温馨愉悦的环境，为用户提供了舒适感和归属感。

设计师：伊斯特法尼亚·贝伦·廷托、维多利亚·阿尔菲里 委托方：拉蒙萨尔达妇产医院 国家：阿根廷

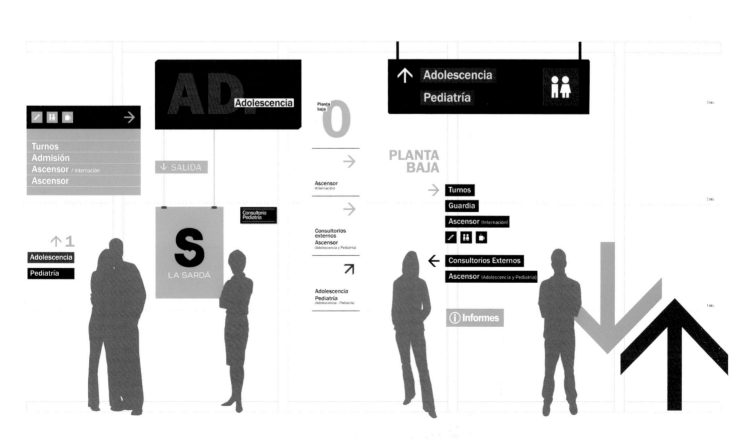

Specialised Hospital

Specialised Hospital

Yoshimoto for Women

Design agency: Mysa Co., Ltd Sign Design
Client: Yoshimoto for Women
Country: Japan

>>

Yoshimoto for Women Hospital commissioned Mysa to design a new signage and identity system. As a comprehensive hospital for women, it provides both traditional obstetrics and medical cosmetology services. The whole design is warm and romantic, perfect for women.

吉本女性医院

吉本女性医院委托 Mysa 为其设计了一套全新的标识及形象系统。该医院是一所综合性的女性医院,既包含传统的产科,又有美容皮肤科。整套设计温馨浪漫,符合女性的特征。

设计机构:Mysa 设计公司 委托方:吉本女性医院
国家:日本

Specialised Hospital

Methodist Women's Centre

Design agency: Gresham, Smith & Partners
Designer: Tim Rucker & Jessica Hill
Client: Methodist Hospital
Country: USA

GS&P'S environmental graphics group provided design and branding services for Methodist Hospital Women's Pavilion in Henderson, Kentucky. Much more than an update, the design of the space blends gracefully, in form and function, with the existing facility-the hallmark of any successful addition. Near completion, it was decided that this newly updated space required a new branding icon. In collaboration with the interior design team, the Graphics Group developed a vocabulary of warm and nurturing tones and materials for the icon design. These tones worked to complement the clean and simple design created for the Methodist Hospital Women's Pavilion brand. The concept for the icon shield was inspired by the floor medallions created for the intersecting hallways. A rich mica material was incorporated into the shield to provide warmth and to emulate the textured finish of the wall treatment. The figurative aluminum ribbon is in reference to the light fixtures used within the newly renovated space; its soft curves are in direct contrast to the stronger lines of the shield. The icon is further enhanced by the use of internal illumination for an inviting glow. This dimensional brand serves as both a warm welcoming feature and an icon for the newly updated space.

卫理公会医院女性医疗中心

GS&P的环境图形设计团队为位于肯塔基州亨德森市的卫理公会医院的女性医疗中心提供了设计和品牌服务。这次空间改造不仅仅是升级，而是将空间设计在形式与功能上与已有设施优雅地融合在一起。在接近完工之时，团队决定为这个新升级的空间打造一个新的品牌图标。平面设计团队与室内设计团队合作开发了一个温暖宜人的图标设计，设计的基调与女性医疗中心简洁的室内设计十分相称。丰富的云母材料被打造成盾牌形状，呈现出温暖的感觉并模仿了墙面的处理。形象的铝制缎带参考了新增的灯具设计，它柔和的曲线与盾牌强烈的线条形成了直接对比。内部照明设计进一步强化了图标的效果，使其散发出友好的光芒。这个新设计的品牌标识既营造了温馨友好的氛围，又是新升级空间的标志。

设计机构：GS&P设计公司 设计师：蒂姆·洛克、杰西卡·希尔 委托方：卫理公会医院 国家：美国

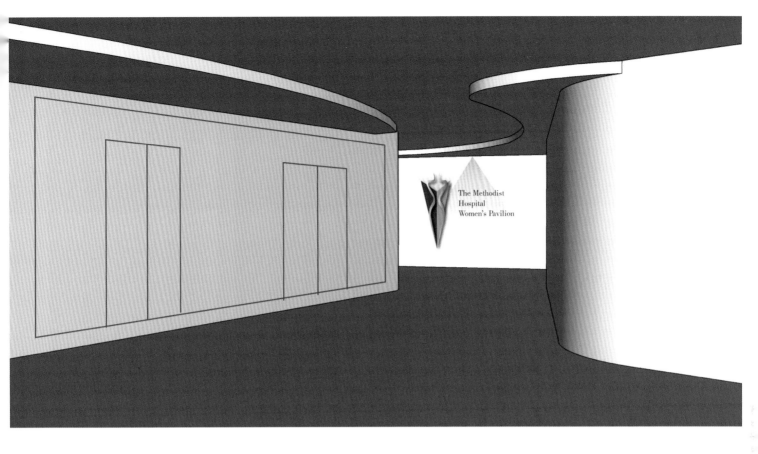

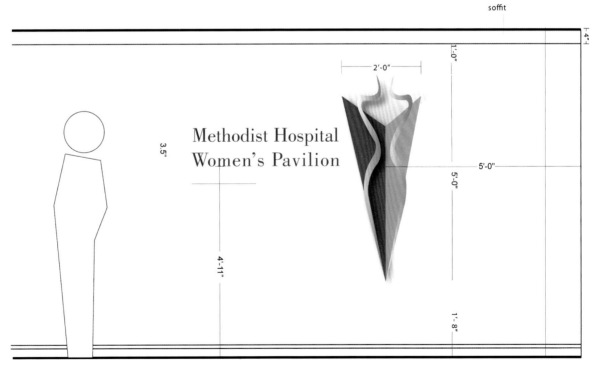

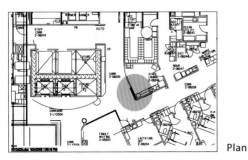

Plan

NOTES

Lettering is 1/4" thick vertical grain. Brushed aluminum, pin mounted.

Brushed aluminum ribbon figure. Mica sconce internally illuminated to glow evenly. Pin mounted light to create "halo" effect on wall surface.

NOTE: Position template for owner approval prior to installation.

Elevation

3/4" = 1'

Perm Perinatal Centre

Design agency: studio gd
Client: Perm Perinatal Centre
Country: Russia

>>

According to the Ministry of Health and Social Development of the Russian Federation, there are 23 perinatal Centres in Russia. Nevertheless, none of them is equipped with a contemporary way finding system, and system of communication. Together with the Perm Centre of Design, the studio has developed the way finding for the Perm Perinatal Centre which allowed turning a huge (28800 square metres) bleak medical building into a friendly space. Thus, the wing for children under 3 years old, was equipped with ABC and corresponding animals on the walls.

彼尔姆孕产中心

根据俄罗斯卫生与社会发展部的数据，俄罗斯目前有 23 家孕产中心。然而，没有一家配有现代化的导视系统。gd 工作室与彼尔姆设计中心共同为彼尔姆孕产中心设计了一套导视系统，将这座 28800 平方米的庞大而荒凉的医疗建筑变成了一个友好的空间。3 岁以下儿童病房楼的墙壁上还添加了 ABC 字母和对应的动物图案。

设计机构：gd 工作室 委托方：彼尔姆妇产中心
国家：俄罗斯

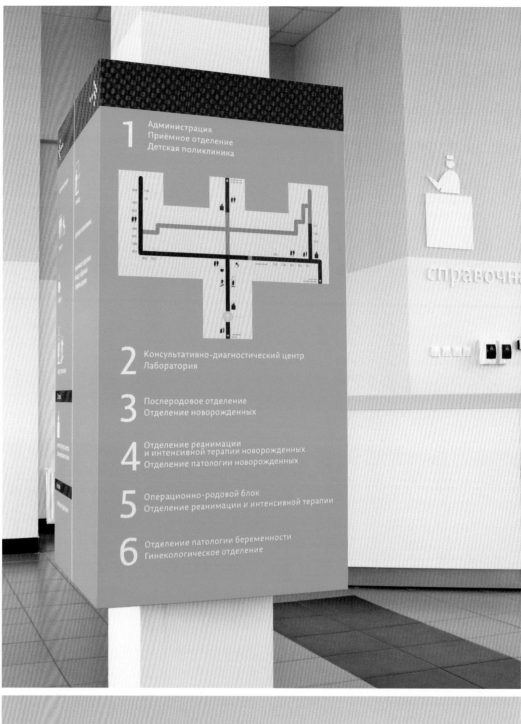

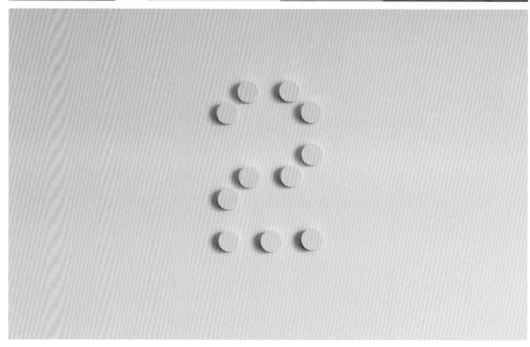

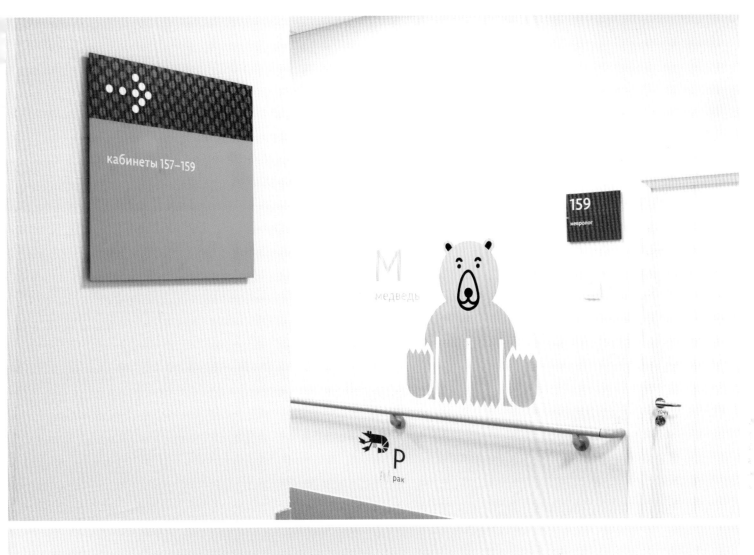
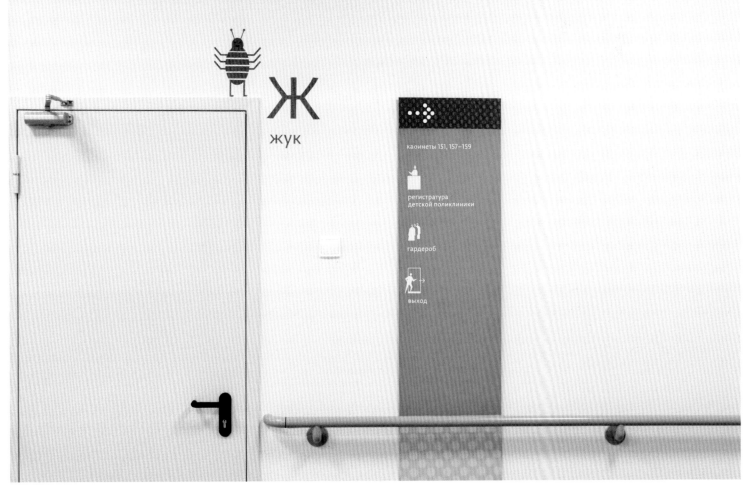

Specialised Hospital

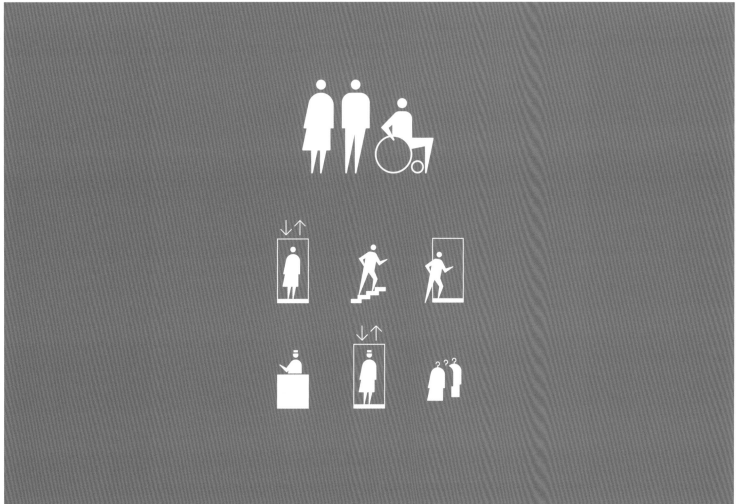

Specialised Hospital

University of California San Diego Sulpizio Cardiovascular Centre and Thornton Hospital Expansion

Design agency: RTKL Associates Inc.
Client: UCSD Facilities Design & Construction
Country: USA

The University of California San Diego Health System came to RTKL with an ambitious undertaking in mind – adding a new, 128,000 SF comprehensive CVC (CVC) to Thornton Hospital on the La Jolla campus. The project has been awarded Modern Healthcare Design Award 2012.

加州大学圣地亚哥心血管中心与桑顿医院扩建

加州大学圣地亚哥分校的医疗系统委托 RTKL 在桑顿医院的拉荷亚院区建造一座约 11890 平方米的心血管中心。项目获得了 2012 现代医疗设计奖。

设计机构：RTKL 建筑事务所 委托方：加州大学圣地亚哥分校设计与建设部 国家：美国

Specialised Hospital

3 **Elevation | Additional Colorways**
Scale: 3"=1'-0

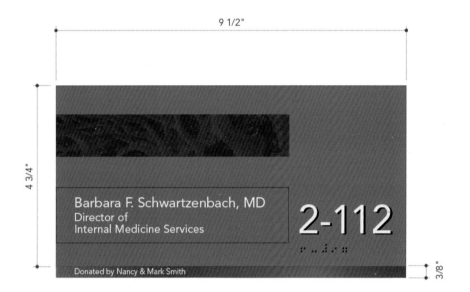
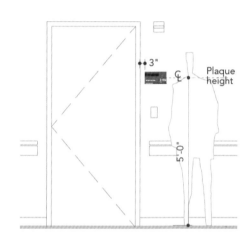

1 **Elevation**
Scale: Half Full Size

2 **Mounting**
Scale: 3/8"=1'-0"

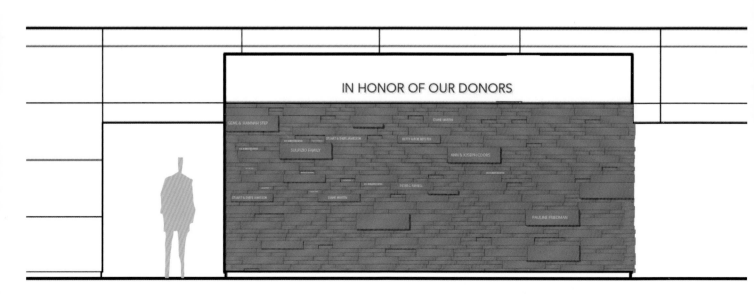

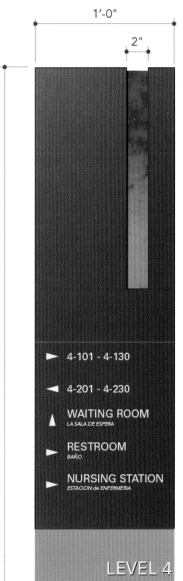

1 Elevation
Scale: 1-1/2"=1'-0

3 Elevation | Additional Colorways
Scale: 3/4"=1'-0

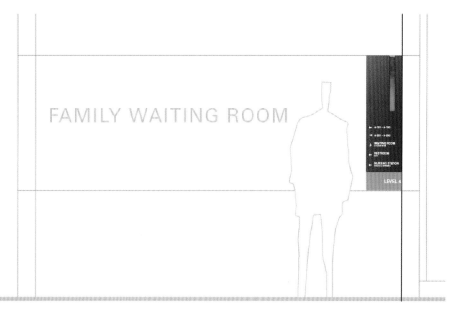

2 Mounting
Scale: 3/8"=1'-0

Specialised Hospital

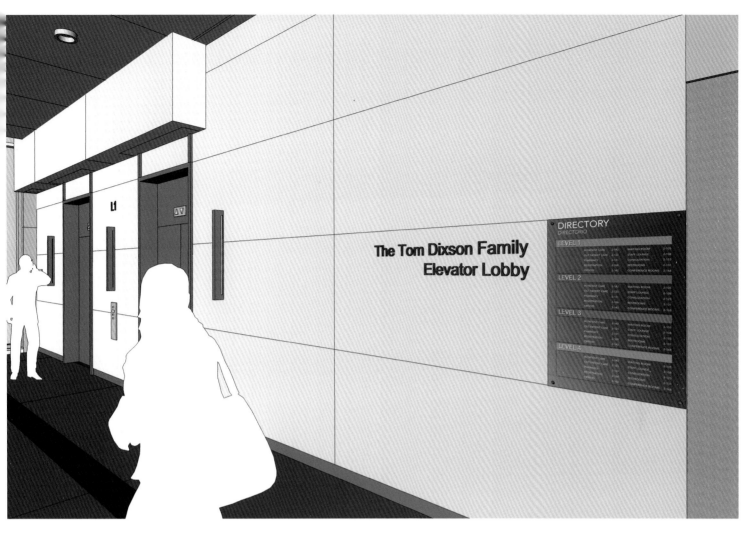
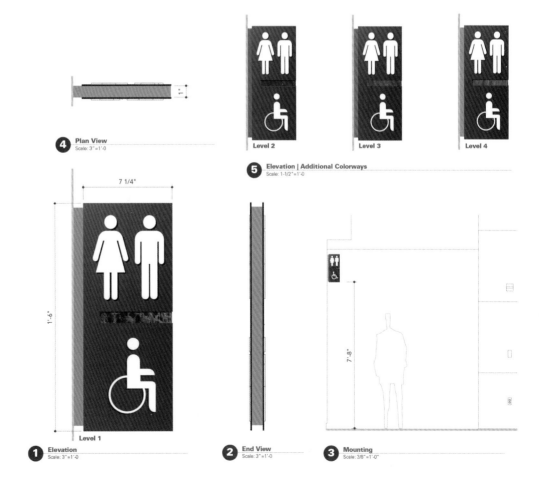

Specialised Hospital

Hospital Veterinari Canis Mallorca

Design agency: Estudi E. Torres Pujol
Designer: Esteve Torres Pujol
Photography: Jose Hevia
Client: Hospital Veterinari Canis Mallorca
Country: Spain

The building is located in an intermediate location between and industrial and a residential area, adjoining the old prison in Palma, nowadays abandoned. The plot is trapezoidal, and the building adapts to it using the maximum surface area allowed, as it was required by the client. Its volumetric architecture is a constant dialogue with the environment, blending the architecture of 'International Style' and traditional rural buildings in Mallorca. The colour world in black and white determines the interior design as well as the sign design.

马略卡宠物医院

建筑位于工业区和住宅区之间，靠近帕尔马被废弃的旧监狱。建筑地块呈梯形，为了最大限度地利用土地，建筑也采用了这种造型。建筑空间与环境形成了对话，将国际风格与马略卡的传统乡村风格融合起来。黑白两色贯穿了室内设计和导视设计。

设计机构：Estudi E. Torres Pujol设计公司 设计师：埃斯特维·托雷斯·普霍尔 摄影：约瑟·埃维亚 委托方：马略卡宠物医院 国家：西班牙

Specialised Hospital

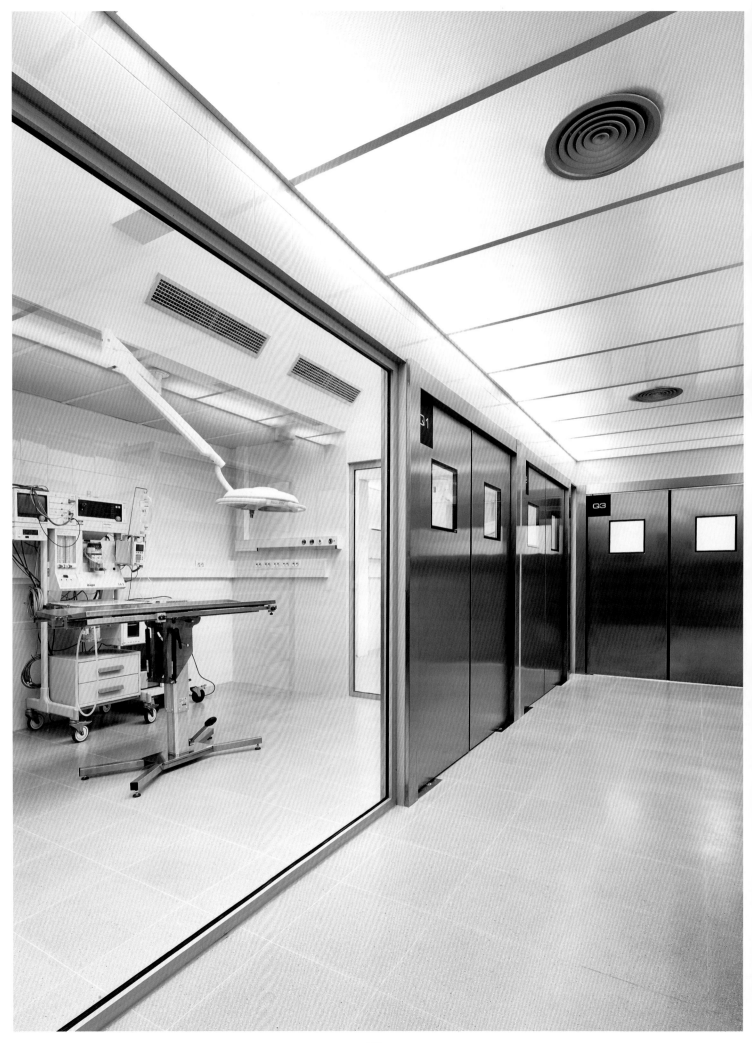

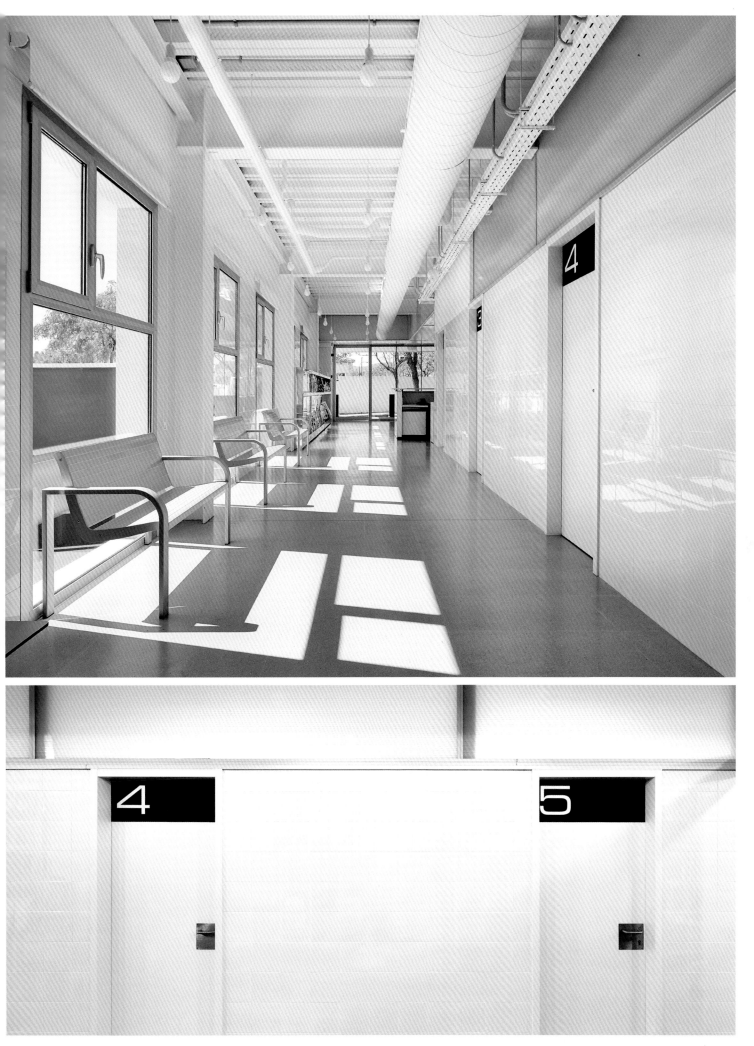

Specialised Hospital

Wittlinger Hahn Stern Radiologie

Design agency: Ippolito Fleitz Group GmbH
Designer: Gunter Fleitz, Peter Ippolito, Axel Knapp, Frank Faßmer, Yuan Peng
Country: Germany

The appearance and directional guidance of the rooms of the "Wittlinger, Hahn, Stern" radiology practice in Schorndorf was developed in the course of architecturally reorganising the practice's interior design. When considering the graphical appearance of the practice, the designers took their orientation from image production in radiology. Highly-specialised technical equipment generates images of the human body that enable medical diagnoses to be made thanks to their precision in detail and penetration. The resulting multi-layered pictorial raster forms the skeletal structure for the graphical appearance. Large-scale, abstract graphics are applied to the walls as central directional guidance points. The practice logo is also derived from this grid of lines. Each of the four treatment areas is encoded via two-tiered door labeling. In the waiting area, a backlit wall displays a blank-and-white cloud formation conveying the pictorial aesthetics of the image reproduction process of medical-technical equipment. This motif is also strongly rastered and continues the basic concept of the graphical appearance on yet another level.

WHS 放射治疗医院

WHS 放射治疗医院的导视设计与室内设计融为一体。在考虑到医院的图形设计时，设计师从射线成片中获得了灵感。专门的技术设备生成人体影像，通过精确的细节呈现和穿透力来辅助医疗诊断。最终形成的多层图像栅格形成了图形外观的骨架结构。大尺寸的抽象图形被应用到墙面上，作为主要的引导标识。医院的标识同样来自于这些线条。四个治疗区都拥有双层门编码。在候诊区，背光式墙面展示了一个白云的造型，体现了医疗技术设备的成像过程。这个图案同样被光栅化，延续了整体设计的图形主题。

设计机构 Ippolito Fleitz Group设计公司 设计师：甘特·弗雷斯、彼得·依普利托、阿克塞尔·克纳普、弗兰克·法富勒米尔、园·彭 国家：德国

Specialised Hospital

AZS – Interior Design and Signage System

Design agency: umsinn.com
Designers: Maximilian Baud, Uwe Strasser
Client: Dr. Kubin
Photography: herruwe.com, Uwe Strasser
Country: Austria

>>

A brand new and modern MRI has been annexed to the X-Ray centre of the AZS (Medical Centre Schallmoos). The interior has been completely redesigned and a signage system has been developed for both the X-ray centre and for the building including the basement and the parking spaces. The centre houses several doctor's offices and keeps apart the X-ray centre in a neutral, independent but strong appearance. To support orientation in the medical centre a symbolic family has been created and a colour concept which integrates each doctor's office independently by its unique colour. In addition the X-ray centre's existing corporate design has been adapted and the symbolic language has been integrated. On top a key-visual of transparent circles that interprets the X-rays completes the interior and defines the staff area and the waiting rooms.

沙尔慕斯医疗中心室内设计与导视系统

沙尔慕斯医疗中心的X射线中心最近新增了一套核磁共振成像系统。X射线中心和整座大楼都进行了全套的室内改造和导视系统再设计。中心内设有若干个医生办公室，呈现出中性、独立而强烈的外观。为了辅助用户在医疗中心内的通行，设计师打造了一套符号元素，并为每个医生办公室打造了一个独特的色彩概念。此外，设计还沿用了X射线中心原有的企业设计和符号语言。

设计机构：umsinn.com 设计公司 设计师：马克西米兰·鲍德、乌维·斯特拉塞尔 委托方：库宾医生 摄影：herruwe.com、乌维·斯特拉塞尔 国家：奥地利

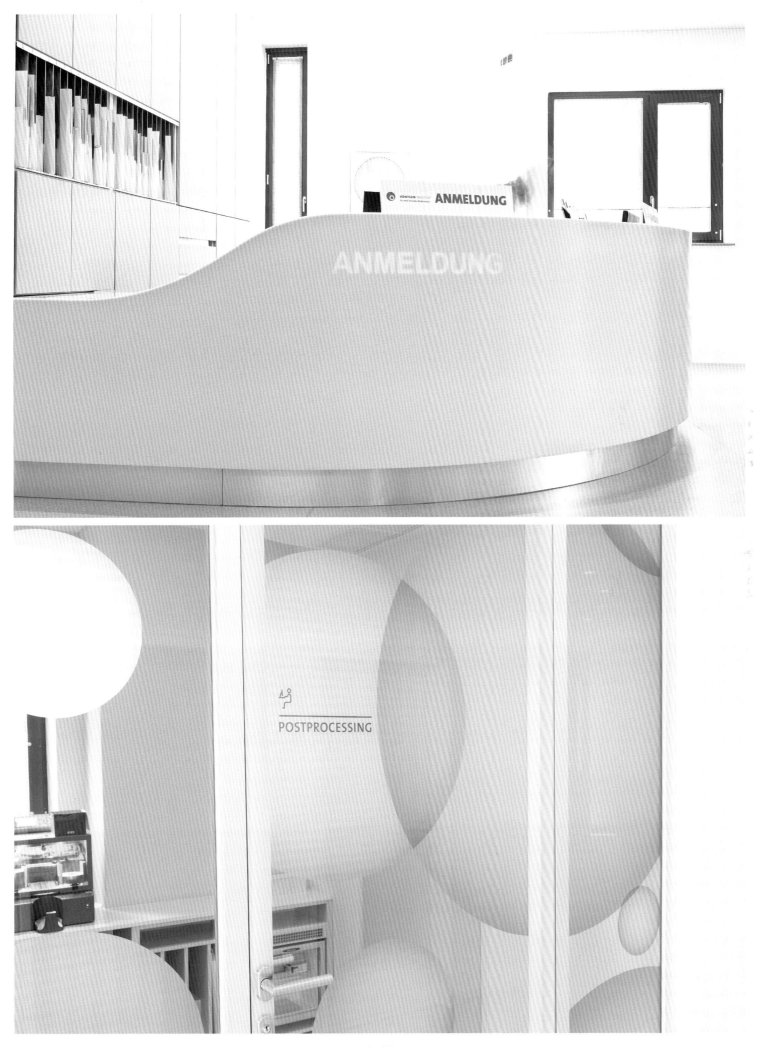

Medical Centre

Medical Centre

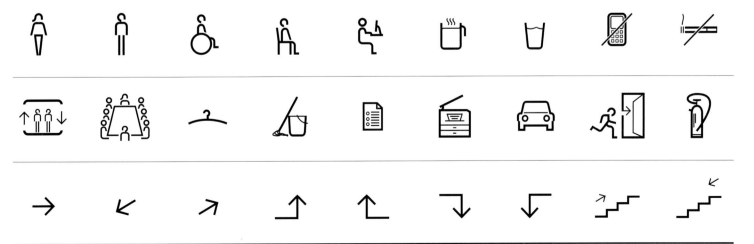

Medical Centre

Baptist Medical Centre

Design agency: Image Resource Group
Client: Baptist Medical Centre
Country: USA

Likened to building a ship in a bottle, the Baptist Wolfson Tower is a Children's and Adult Hospital that is part of the Baptist Health System and Wolfson Children's Hospital in Jacksonville, Florida. The new tower is an extension of the existing hospital facilities in Downtown and presented some incredibly complex and tight constraints within an urban site. There were functional challenges as well with regards to the integrated pediatric and adult populations within one building. The pediatric hospital serves children within the pan-handle as well as South Georgia. Each floor of the pediatric hospital consists of a carefully selected nature motif, including rivers, forests, and grasslands that reflect the nearby natural surroundings. Each room has floor to ceiling glass windows and its own animal theme. All of the pediatric rooms are family suites and 100 square feet bigger than their old rooms. The full bathrooms, refrigerators, safes and beds make it feel more like moving into a hotel than a hospital. The upper floors contain the adult beds. Unlike in the previous hospital, all of the rooms will be private. Each room is 300 square feet, twice the size of the rooms in the old building, which was built in 1955. The adult floor contains bridge themes in muted colours with beautiful views of the St. Johns River.

浸礼会医疗中心

浸礼会沃尔夫森医疗中心是一所隶属于浸礼会医疗系统和沃尔夫森儿童医院的儿童与成人综合医院。新建的医院楼是对原有医院设施的扩建，深受城市内部的空间限制。设计的功能性挑战主要是如何将儿科与成人科结合在同一座楼内。儿科中心主要为南佐治亚地区的儿童提供医疗服务，每层楼都配有精选的自然图片，包括河流、森林、草地等反映附近自然景观的图像。每个房间都配有落地窗和动物主题。所有儿科病房均为家庭套房，比之前的病房更大。浴室、冰箱、保险柜和床位让这里看起来更像是酒店而不是医院。建筑上层是成人病房，与之前不同，所有病房都是独立的。每个病房都有近28平方米，是之前的两倍。成人楼层以色彩柔和的大桥为主题，配以圣琼斯河的美景。

设计机构 Image Resource Group设计公司 委托方：浸礼会医疗中心 国家：美国

Medical Centre

Medical Centre

Health Care REIT

Design agency: Gresham, Smith & Partners
Designer: Tim Rucker & Mike Summers
Client: Health Care REIT
Country: USA

GS&P provided environmental graphic design services for Health Care REIT's (Real Estate Investment Trust) headquarters, a two storey building in Brentwood, Tennessee. Contemporary and creative solutions were employed in the designs to reflect the clean and simple brand of the company. With the potential of multiple visitors, investors and medical professional visiting the new office space, it was important that the space be considered a showpiece and an example of Health Care REIT's properties-reflecting the creative culture of their business and employees. Design goals and concepts for the graphics were created around the following three principles: collaboration, image/brand and representation of Health Care REIT as a company the creative staff and the properties in which Health Care REIT develops, invests and manages. The environmental graphics were designed to complement the materials used in the renovation of the building, to communicate Health Care REIT's corporate culture and create functional modern designs for the building's signage. Design decisions were first based on function, by encouraging collaboration between employees and departments. The designs were enhanced by the use of both colour and translucent materials, thereby linking the multiple team areas together throughout the newly renovated space. By use of these branded graphics and a consistency between interior material finishes and furniture, the visitor and employee experience is that of a dynamic, interactive and inviting space.

地产投资信托医疗中心

GS&P设计公司为地产信托医疗中心的总部提供了环境图形设计服务，这座大楼位于田纳西州的布伦特伍德市，由两层楼构成。设计运用了现代创新方案来反映公司品牌的简洁利落之感。由于未来将由各种访客、投资者和医疗专家前来到访，空间的整体感和企业文化的体现显得至关重要。图形设计的目标和概念围绕着三个主题展开：合作、图像/品牌、"地产信托医疗公司是由创意工作人员所开发、投资并管理的公司"。环境图形设计与建筑翻修所使用的材料相匹配，表现了公司的企业文化，为建筑打造了现代而实用的建筑标识。设计以功能为基础，鼓励员工与部门间的合作。色彩与半透明材料的使用强化了设计的效果，将不同的团队区域连接起来。品牌图形以及统一的室内设计材料和家具让访客和员工的体验更加完整，形成了一个活跃、互动的友好空间。

设计机构：GS&P设计公司 设计师：蒂姆·洛克、麦克·萨默斯 委托方：地产信托医疗中心 国家：美国

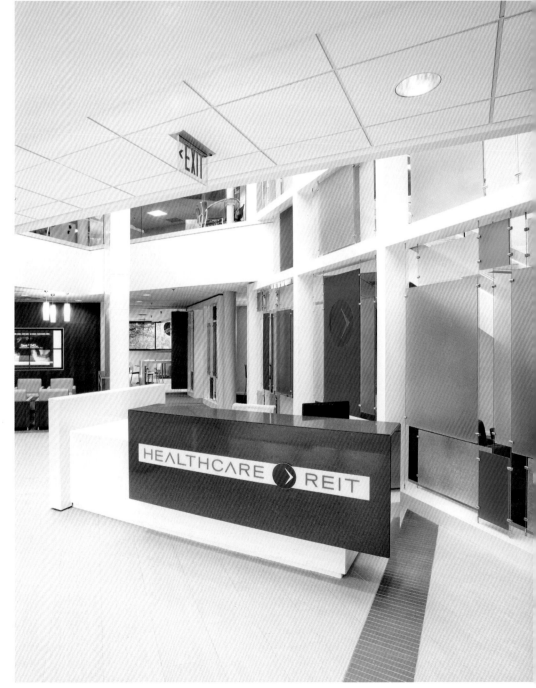

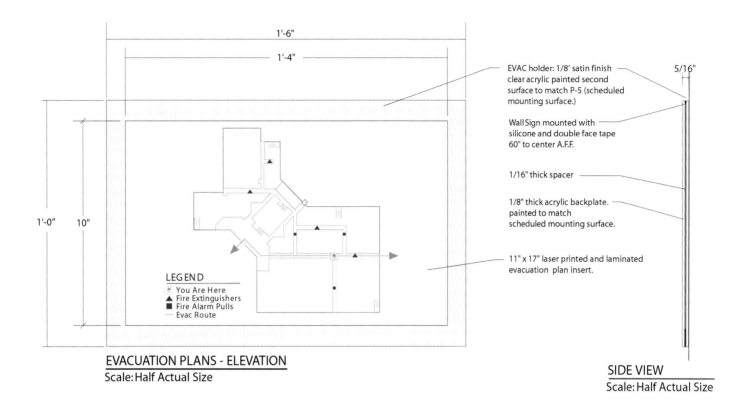

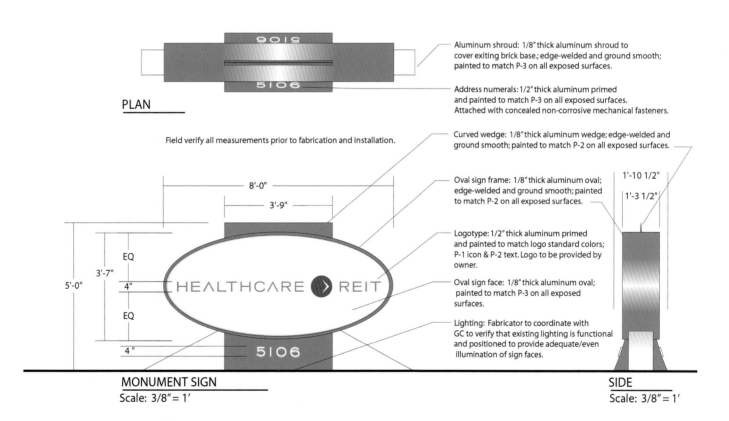

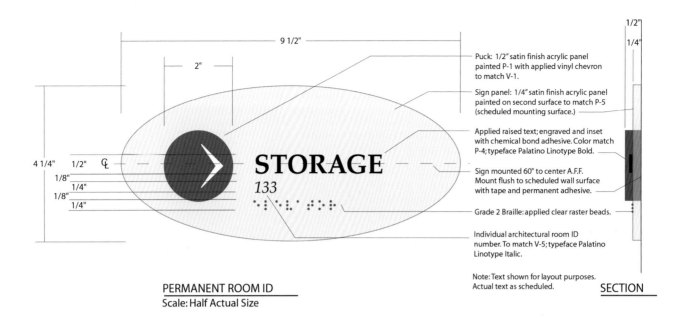

PERMANENT ROOM ID
Scale: Half Actual Size

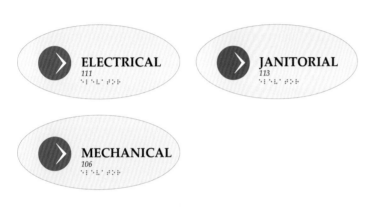

ALTERNATE LAYOUTS
Scale: 3" = 1'-0"

PERMANENT ROOM - typical mounting

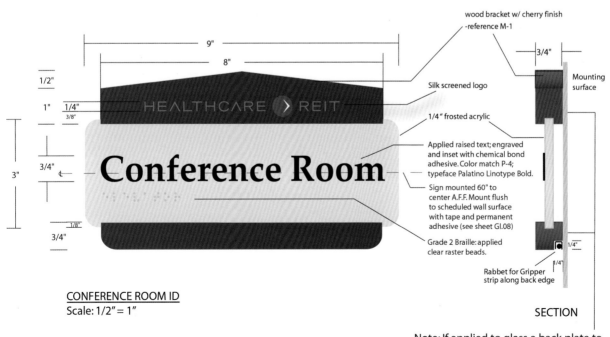

CONFERENCE ROOM ID
Scale: 1/2" = 1"

Note: If applied to glass a back plate to conceal attachment is required.

Medical Centre

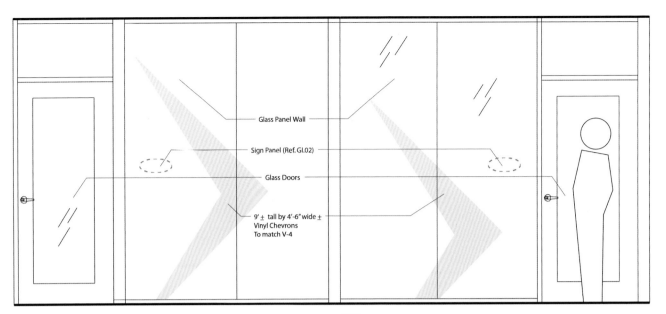

VINYL VISION SCREEN - typical

NOTE:
Applied vinyl sChevrons: To match V-4; applied second surface to glass walls; Field verify locations, quantity, and dimensions with owner prior to fabrication; Position for owner approval prior to final installation.

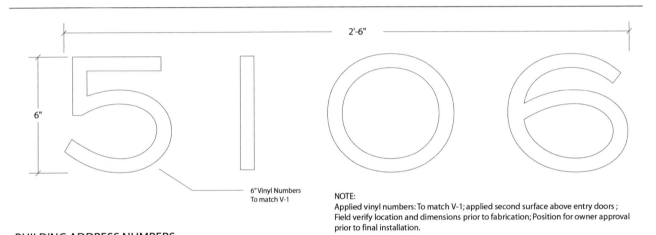

BUILDING ADDRESS NUMBERS

NOTE:
Applied vinyl numbers: To match V-1; applied second surface above entry doors; Field verify location and dimensions prior to fabrication; Position for owner approval prior to final installation.

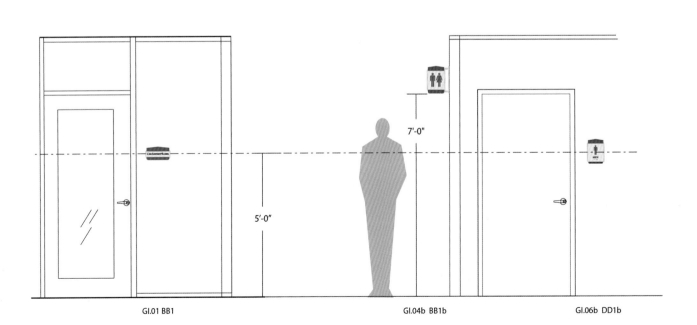

Gl.01 BB1   Gl.04b BB1b   Gl.06b DD1b

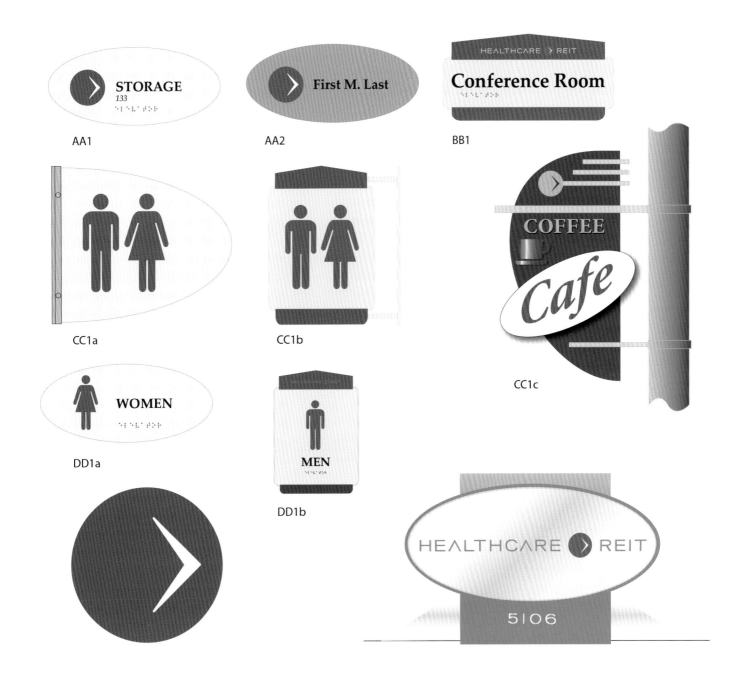

Palatino Linotype Bold

Aa Bb Cc Dd Ee Ff Gg Hh Ii Jj Kk Ll Mm Nn Oo Pp Qq Rr Ss Tt Uu Vv Ww Xx Yy Zz 0123456789

Palatino Linotype Italic

Aa Bb Cc Dd Ee Ff Gg Hh Ii Jj Kk Ll Mm Nn Oo Pp Qq Rr Ss Tt Uu Vv Ww Xx Yy Zz 0123456789

Symbols & Logos

S1 - In Case of Fire S2 - Stairs S3 - Elevator S6 - Men S7 - Women S8 - Handicap S9 - Coffee Cup

Medical Centre

Kaiser Permanente Southwood Comprehensive Medical Centre

Design agency: Gresham, Smith & Partners
Designer: Tim Rucker
Client: Kaiser Permamente
Country: USA

Guided by Kaiser Foundation Health Plan's (KFHP) design guidelines, GS&P provided a comprehensive building design, site layout and phasing plans for the new Kaiser Permanente Southwood Comprehensive Medical Centre. While renovations were performed throughout the entire Southwood facility, the most significant renovations took place on the first floor, where GS&P expanded the primary care and pediatric clinics and added a new central check-in area. The design team implemented Lean design principles, reorganising medical assistant work spaces into a centralised area to allow for efficient staff movement. The expansion component comprises additional specialty clinics, an acute care Centre and a procedure suite. By integrating these functions into one facility, the medical Centre significantly expanded its existing departments and services, resulting in the most complete patient care possible in an outpatient setting. To connect the new facility to the original medical Centre, GS&P designed a two-storey glass walkway inspired by KFHP's "Thrive" campaign.

凯萨永久医疗集团索思伍德综合医疗中心

在凯萨基金会健康计划的设计指导下，GS&P 为新建的凯萨永久医疗集团索思伍德综合医疗中心提供了综合建筑设计、场地规划以及分期规划。整个医疗中心都经历了翻修，其中最明显的翻修工作在二楼。GS&P 拓展了初级护理和儿科门诊，并新增了一个中央登记区。设计团队实施了精益设计原则，重新将医疗辅助工作空间集中起来，提高了人员的移动效率。扩建包含新增的专科门诊、急救中心和处置套房。这些新设施的加入让医疗中心的科室和服务得到了大幅拓展，为患者提供了完整的医疗服务。为了将新设施与原有的医疗中心连接起来，GS&P 设计了一个双层玻璃通道，该设计从凯萨基金会健康计划的"繁荣活动"中获取了灵感。

设计机构：GS&P 设计公司 设计师：蒂姆·洛克 委托方：凯萨永久医疗集团 国家：美国

Medical Centre

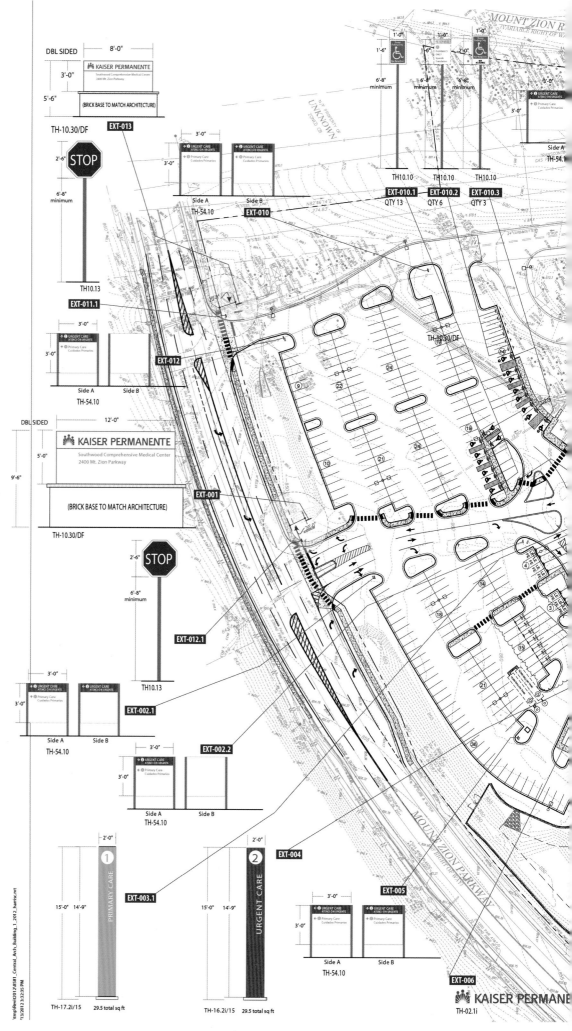

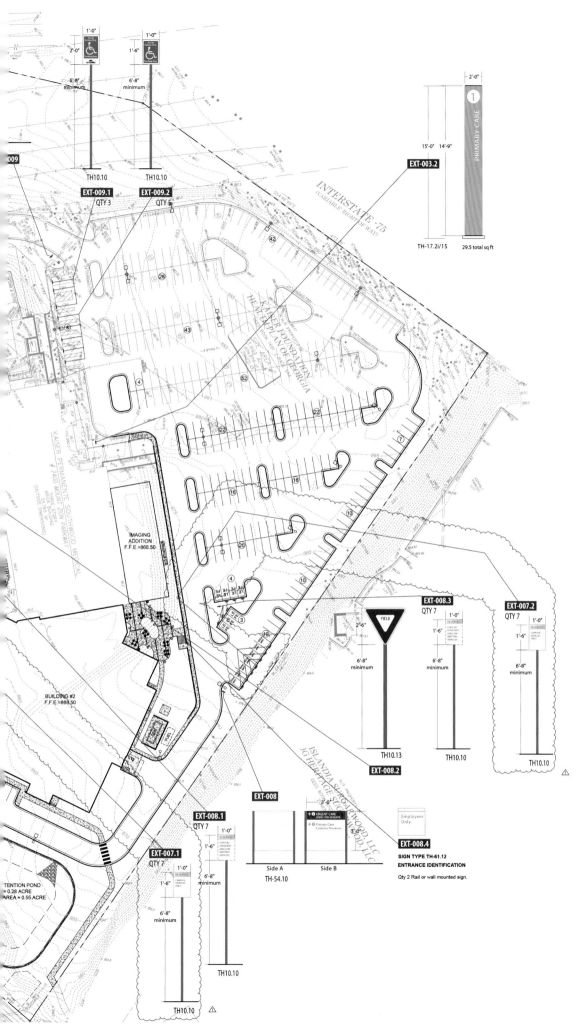

Medical Centre

Medical Centre

Lentz Public Health Centre

Design agency: Gresham, Smith and Partners
Designer: Mike Summers & Lauren Comet
Photography: Chad Mellon & Gresham, Smith and Partners
Client: Metro Public Health Department
Country: USA

>>

Guiding Principles for the graphic design of Lentz Public Health Centre was to "enhance services through hospitality, efficiency, equality, accessibility and flexibility". Working closely with the architectural and interior design teams, GS&P's environmental graphics group created interior and exterior signage and wayfinding solutions that integrated with the architectural language of the building and worked in conjunction with the architectural colour palette & design direction. Specific interior wall colours for each floor were carried over to interior signage to help create intuitive wayfinding throughout the 3 storey building. Iconography was created from departmental logos and used as large format confirmation graphics at the public entrances to each department. A circular theme was developed for the interior space branding and was utilised for vision screening at conference rooms as well as directory and directional information in public corridors. Wall graphics themed to individual departments were also created for several of the waiting areas.

伦茨公共健康中心

伦茨公共健康中心的导视设计指导原则是"通过热情、高效、平等、亲和、灵活的设计提升服务品质"。GS&P的环境图形设计团队与建筑和室内设计团队紧密合作，打造了室内外导视策略，使其与建筑语言融为一体，与建筑色彩搭配和设计方向相一致。每层楼的内墙被赋予了不同的色彩，在三层楼之间形成了直观的区别，有助于室内导视的设计。设计师为各个部门都设计了图表，并将其应用在每个部门的公共入口处，以大幅图形的形式呈现出来。室内空间以圆形为主题，圆形图案被应用在会议室的屏风以及公共走廊的引导信息上。一些候诊区还添加了与各部门相应的壁画。

设计机构：GS&P设计公司 设计师：麦克·萨默斯、劳伦·科梅 摄影：查德·梅隆、GS&P设计公司 委托方：城市公共健康部 国家：美国

Medical Centre

Osa Integrated Centre

Design agency: QUA DESIGN style
Designer: Yuichiro Tanaka
Client: Niimi city
Country: Japan

Osa Integrated Centre accommodates the facilities to plan promotion of art, the culture of local inhabitants and the life study and an increase of health, the welfare. The signage system uses typography and concise icon design to illustrate the information visually, enabling visitors to get the informative content quickly and clearly.

Osa 综合中心

Osa 综合中心内设置着本地艺术文化宣传部门、生活研究部门和医疗保健福利部门。这套导视通过文字设计和简洁的图标设计直观的展现了所要传达的信息，使受众能够迅速清晰的了解导视设计中体现的信息内容。

设计机构：QUA DESIGN 设计公司 设计师：田中裕一郎 委托方：新见市 国家：日本

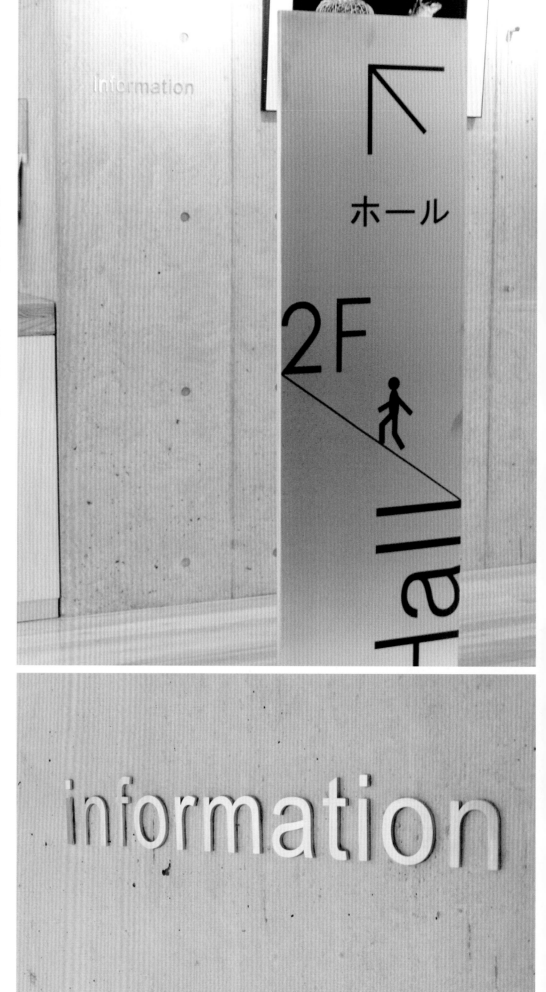

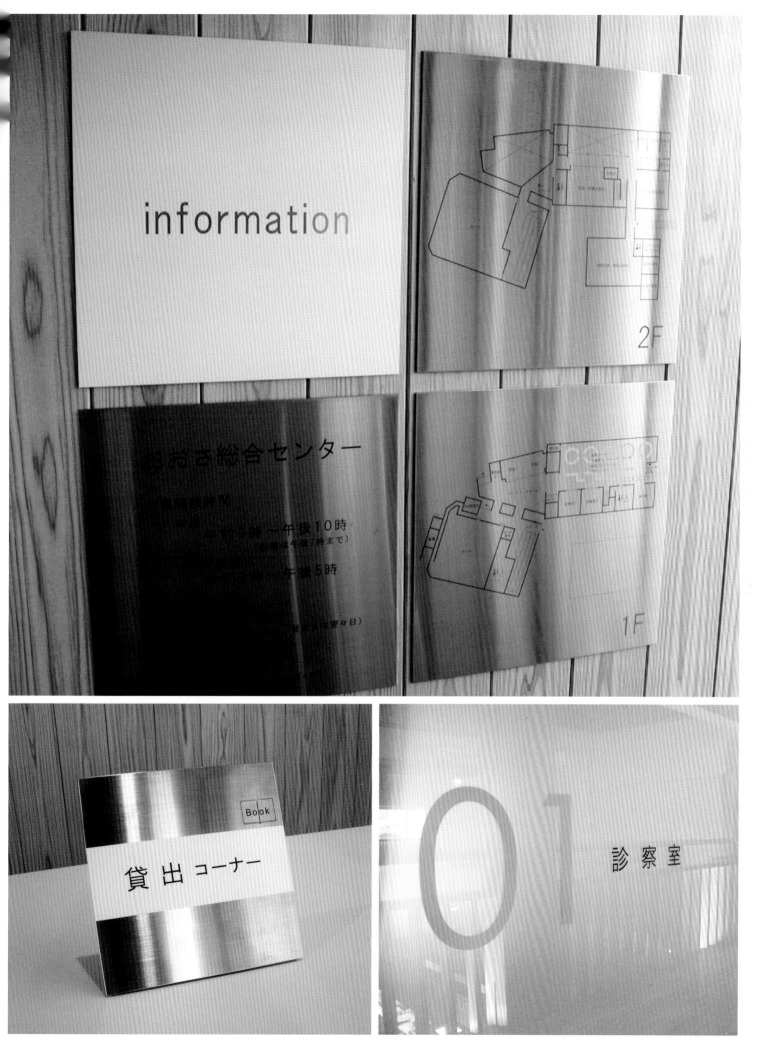

Medical Centre

San Antonio Military Medical Centre

Design agency: RTKL Associates Inc.
Photography: Dave Whitcomb, RTKL
Client: US Army Corps of Engineers
Country: USA

San Antonio Military Medical Centre opened in September 2011 as the nation's largest Defense Department (DOD) inpatient facility. An information-driven system is based on a user-centric approach to wayfinding design. The team utilised on-site prototyping and user interviews to create a "user-centric" design solution that consists of a simple "visitor and patient pathway" that connects all the public vertical circulation.

圣安东尼奥军事医疗中心

圣安东尼奥军事医疗中心于2011年9月正式开放，是美国最大的国防部住院机构。医疗中心的信息系统采用了以用户为中心的引路设计。设计团队利用现场原型设计和用户访谈来打造了一个"以用户为中心"的设计方案，由简单的"访客与患者通道"连接了各个楼层之间的垂直交通。

设计机构：RTKL建筑事务所 摄影：戴夫·惠特科姆、RTKL 委托方：美国陆军工程部队 国家：美国

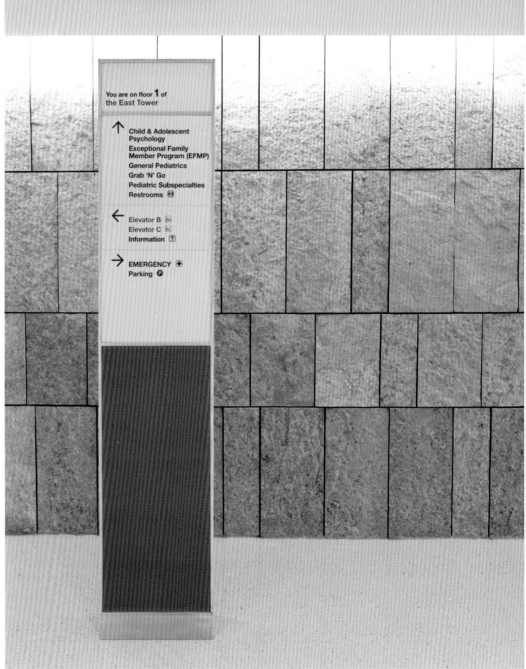

Medical Centre

Medical Centre

Mater Private Hospital

Design agency: Jennings design studio
Photography: Matthew Thompson
Client: Mater Private Hospital
Country: UK

>>

This is a collaborative project to modernise the environment of the new Heart Centre. Along with the branding and wayfinding designed by CO/DE the aim was to give the space an approachable, contemporary feel appropriate to a modern medical facility.

梅特私立医院

项目为医院新建的心脏中心打造了环境图形设计，由 CO/DE 设计品牌形象和引路系统。项目的目标是给予空间亲切、现代的感觉，与该机构作为现代医疗设施的身份相匹配。

设计机构：Jennings 设计工作室 摄影：马修·汤普森 委托方：梅特私立医院 国家：英国

Medical Centre

Isar Medical Centre

Design agency: Gourdin & Müller
Designer: Nathanaël Gourdin, Katy Müller, Katharina Seitz
Client: Isar Klinik II AG
Country: Germany

Cheerful colours and rounded forms – with regard to both typography and the signs themselves – are the underlying characteristics of the design concept. The overlapping of grid areas results in multi-layered graphics that pay visual tribute to medical expertise on the one hand and convey transparency and purity on the other.

伊萨尔医疗中心

活泼的色彩和圆角造型是设计的主要特色,它们不仅被应用在字体上,还被应用在标识自身上。网格区域的重叠形成了多层图形,一方面在视觉上突出了医疗技术,一方面又体现了通透感和纯粹感。

设计机构:Gourdin & Müller 设计公司 设计师:纳萨奈尔·戈尔丁、凯蒂·穆勒、卡特琳娜·塞茨 委托方:伊萨尔医疗公司 国家:德国

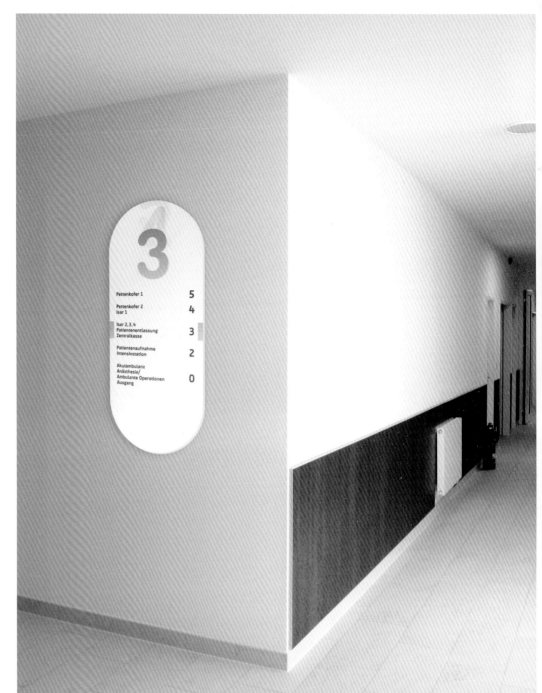

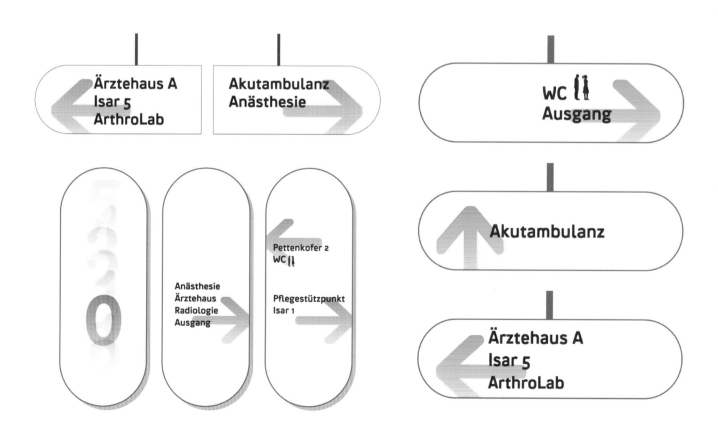

Medical Centre

UQ Oral Health

Design agency: Dotdash
Designer: Heath Pedrola
Photography: Angus Martin
Client: Cox Rayner Architects
Country: Australia

In order to achieve a strong design intent, Dotdash was engaged to extend the architects vision, and create a family of signage that works in unity with the built form. Dotdash has implemented a wayfinding strategy which functions to clearly navigate the Centre's students and patients throughout its many facilities, whilst also reinforcing a broader design vision.

昆士兰大学口腔健康中心

为了实现强烈的设计意图,Dotdash 拓展了建筑师的愿景并打造了一系列与建筑形式相匹配的导视元素。Dotdash 所完成的导视策略能够为学生和患者提供清晰的导视服务,同时也能强化更广义的设计愿景。

设计机构:Dotdash 设计公司 设计师:希斯·佩德罗拉 摄影:安格斯·马丁 委托方:Cox Rayner 建筑事务所 国家:澳大利亚

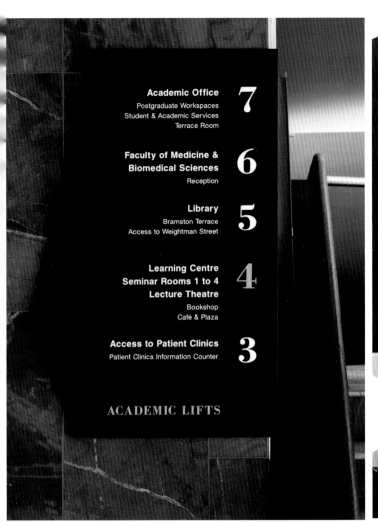

Medical Centre

North Richmond Community Health Centre

Design agency: ID/Lab
Photography: ID/Lab
Client: Aurecon
Country: Australia

ID/Lab was approached by Aurecon to develop a wayshowing strategy and associated environmental graphic design for the North Richmond Community Health (NRCH) project. ID/Lab decided to base the signage design on the architectural language used by Lyons Architects in the design of the furniture and the building. Using routered plywood panels for most of the signage hardware was both practical and functional. The demographics of NRCH's clients called for a wayfinding and an information system that worked for a large culturally and linguistically diverse group, many of whom had complex care needs. Part of the wayfinding solution are specially designed pictogram storyboards that inform non-English speakers of some rather intricate instructions. These pictograms were extensively tested with the users before being implemented.

北里士满社区健康中心

ID/Lab 设计公司受 Aurecon 委托为北里士满社区健康项目设计一套引路策略和相关的环境图形。ID/Lab 决定将 Lyons 建筑事务所在建筑和家具设计中所运用的建筑语言作为导视设计的基础。设计师用胶合板作为大多数导视标识的硬件材料，简单实用。北里士满社区健康项目的受众人群要求导视信息系统必须满足各种不同文化、不同语言的群体的需求，其中一些人还有复杂的护理需求。导视策略的一部分专门设计了象形图板来帮助非英语人士。这些象形图在正式安装之前已经经过了广泛的用户测试。

设计机构：ID/Lab 设计公司　摄影：ID/Lab 设计公司　委托方：Aurecon 顾问公司　国家：澳大利亚

Medical Centre

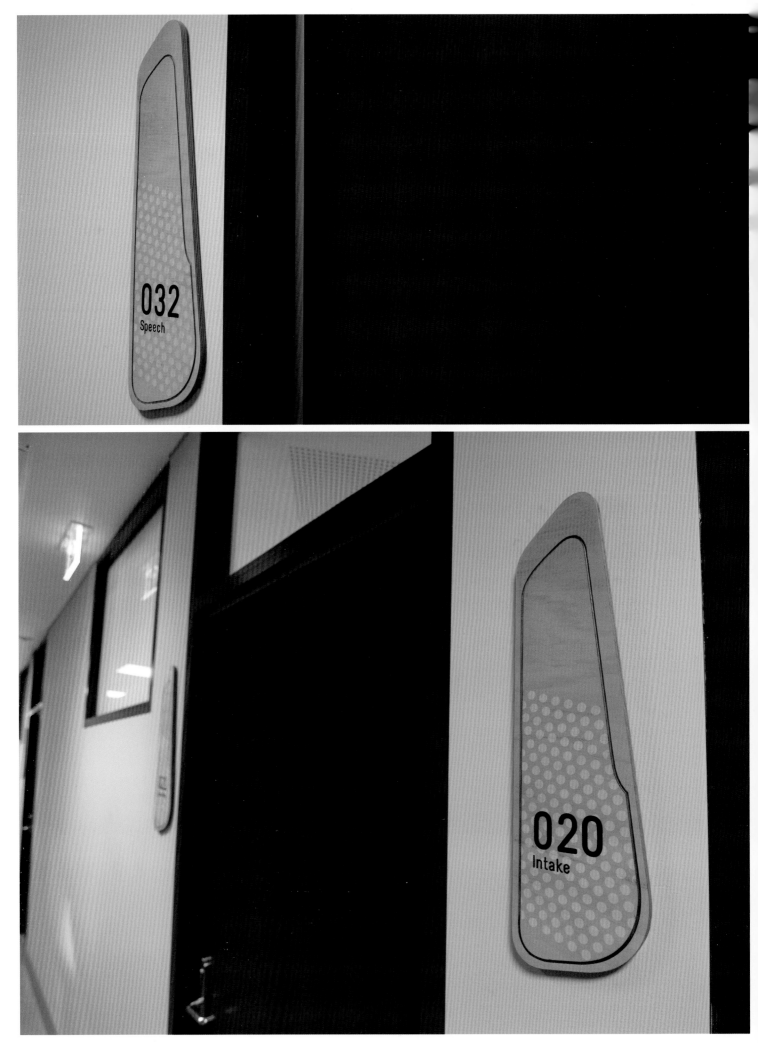

Medical Centre

The South Australian Health and Medical Research Institute

Design agency: ID/Lab
Photography: ID/Lab
Client: Woods Bagot
Country: Australia

The South Australian Health and Medical Research Institute in Adelaide (SAHMRI) houses 670 researchers over nine fully flexible wet and dry laboratory modules, all designed to PC2 standards, which sit alongside vivariums, a cyclotron, and associated public areas. Woods Bagot engaged ID/Lab to develop a complete signage package and wayfinding strategy for this iconic building. The building has a unique triangulated dia-grid façade, inspired by the skin of a pine cone. Driven to make the signs feel like they are part of the architectural fabric (and not just screwed on top of it), ID/Lab developed the design together with the architects and interior designers from Woods Bagot. This allowed them to have space for signage elements cut out of the joinery, have illumination incorporated into the joinery, and have large sized graphics painted onto concrete walls in the carpark. For the exterior sign elements, SAHMRI chose a sculptural expression of the building footprint, which the designers then clustered to create a landmark identifying the path to the main entrance. In the interior, the use of strong coloured, integrated elements and the clever window graphics make the end result feel as if it is the work of the architects.

南澳大利亚健康与医疗研究院

位于阿德莱德的南澳大利亚健康与医疗研究院拥有670名研究员，研究员的9个灵活的干湿试验舱内全部按照PC2标准设计，同时还配有植物园、回旋加速器及配套公共区域。Woods Bagot建筑事务所委托ID/Lab为这座地标性建筑开发整套的导视及导航策略。建筑拥有独特的三角网格立面，其设计灵感来自于松果的表皮。为了让标识与建筑融为一体，ID/Lab与来自Woods Bagot的建筑师和设计师共同开发了导视设计。各方面的合作为导视元素提供了空间，使其融入了照明设施，并且让大尺寸的图形直接喷涂到停车场的混凝土墙上。在室外标识设计中，研究院选择了一种富有雕塑感的表现形式，设计师将它们聚集起来形成了地标，标识出建筑的正门。在室内标识设计中，色彩强烈的整合元素和智能的窗口图形让标识与建筑无缝融合，就像是建筑师亲自设计的一样。

设计机构：ID/Lab设计公司 摄影：ID/Lab设计公司 委托方：Woods Bagot建筑事务所 国家：澳大利亚

Medical Centre

Medical Centre

Willow Creek Care Centre

Design agency: Alex Daye Design
Designer: Alex Daye
Photography: Alex Daye
Client: Willow Creek Community Church
Country: USA

The Care Centre project included the naming, branding, interior elements, and wayfinding of a ten million dollar Care Centre serving the local community with food, clothing, medical, dental, vision care, employment, education, transportation, legal and financial resources. Over 17,000 individual families are served by the Care Centre annually.

柳树溪护理中心

护理中心项目包括命名、品牌推广、室内设计元素和引路设计。这座价值1000万美元的护理中心为当地社区提供食品、服装、医疗、口腔、视力保健、就业、教育、交通和财政资源服务,每年有17000多个独立家庭从中受惠。

设计机构:Alex Daye设计公司 委托方:亚历克斯·达耶 摄影:亚历克斯·达耶 委托方:柳树溪社区教会 国家:美国

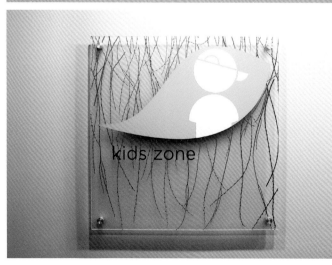

Medical Centre

Kyoto University Hospital / Clinical Research Centre for Medical Equipment Development

Design agency: emmanuelle moureaux
Photography: Daisuke Shima / Nacasa & Partners Inc.
Country: Japan

The design expresses the invisible "threads" that connect each of these different thoughts to one another, just like how threads are spun together to create a strong, supple fabric. Specifically, ito (Japanese for "thread") is used as a motif that would bridge the first and second floors of this research Centre, designing a space that came together in a single, massive flow. Just like how new possibilities emerge out of encounters between people, a spectrum of different colours appear at the junctions between threads, creating chromatic combinations that resemble landscapes: field green, sky blue, light cherry pink, snow white, dusky orange, and white horizons.

京都大学医院 / 医疗设备开发临床研究中心

设计用无形的"线"将不同的思想连接起来,就像是用线织成了一张灵活的大网。设计师特别用"线"作为图形元素,将研究中心的二楼和三楼连接起来,形成一个单一、流畅的空间。就像是人们邂逅的机会一样,两根线在接头处呈现出各种不同的色彩,形成了与风景相类似的色彩组合:草地绿、天空蓝、樱花粉、雪白、黄昏橙和白色的地平线。

设计机构:emmanuelle moureaux 设计公司 摄影:岛大辅、Nacasa & Partners Inc. 国家:日本

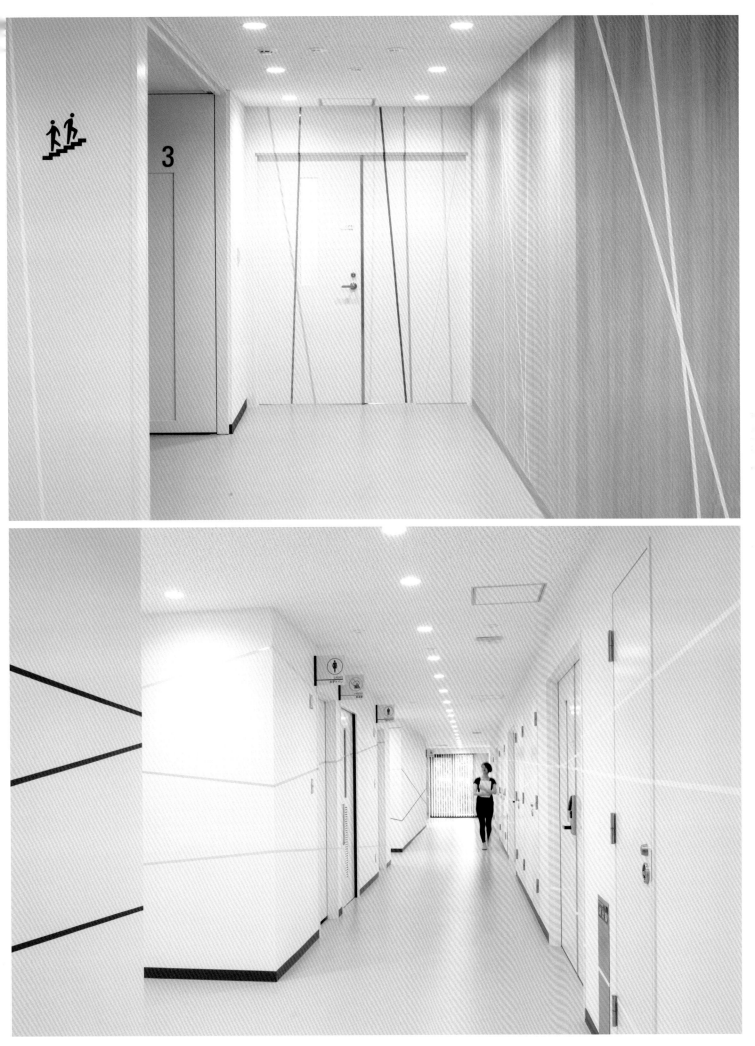

Medical Centre

Sendagrup Medical Centre

Design agency: PAUZARQ arquitectos
Designers: Felipe Aurtenetxe & Elena Usabiaga
Photography: Xabier Aldazabal
Client: Sendagrup Médicos Asociados
Country: Spain

Leaning the consulting rooms beside the facades, the central space was allocated to host the service rooms. These ones are inside an illuminated box that provides enough brightness to the circulation areas. There is a contrast between the two building systems used to rise the partition walls of the consulting rooms. Some of them are heavy and dark, made of marble, whereas the others are made of glass that enables the light to go inside.

森达格拉普医疗中心

诊疗室被设在靠外墙的一侧，而中央空间则用作服务室。交通区域享有充足的采光。在诊疗室的隔断设计中，建筑采用了两个对比系统。一些墙壁厚重深沉，由大理石制成；另一些则由玻璃制成，能让光线进入室内。

设计机构：PAUZARQ建筑事务所 设计师：菲利普·奥特内策、艾琳娜·乌萨维亚加 摄影：伊克萨贝尔·阿尔达萨巴尔 委托方：森达格拉普医疗公司 国家：西班牙

Medical Centre

Cibeles – Health Clinic and Residential Centre for the Elderly

Design agency: Estudi Conrad Torras
Designer: Conrad Torras, Gerard Gris & Eva Pérez
Client: Barcelona Town Council
Photography: Conrad Torras
Country: Spain

>>

Signage for the building of a former nightclub called "La Cibeles", located at 363 Còrsega Street, Barcelona. This building now includes accommodation for the elderly, a Public Health Clinic and car park. Three graphic tools were used to enable understanding and movement through the building. Firstly, printing was used with clear straight lines. Secondly, pictograms were created that turn into initials for each room. Finally, in terms of colour, each floor has been given a single, clearly defined colour (without using shades of grey) so that it might be easily remembered. These three graphic guidelines facilitate movement throughout the building because simplicity translates into comprehension.

库柏勒老年健康诊所与居住中心

建筑的前身是一家名为"库柏勒"的夜总会，位于巴塞罗那科西嘉街363号。目前它集老年公寓、公共健康诊所和停车场设施于一身。为了帮助人们了解建筑并在其中移动，设计师采用了三种图形工具。一是简单的直线印刷术；二是象征着每个房间首字母的象形图标；最后是每个楼层不同的色彩（不使用灰色调）。这三种图形引导能够辅助人们在楼内的运动，简洁明晰。

设计机构：Estudi Conrad Torras设计公司 设计师：康拉德·图拉斯、杰拉德·格里斯、伊娃·佩雷斯 委托方：巴塞罗那市议会 摄影：康拉德·图拉斯 国家：西班牙

Clinic

Clinic

Furukawa Clinic

Design agency: Mysa Co., Ltd Sign Design
Designer: Nariyoshi Kadota
Photography: Takafumi Yamada
Client: Furukawa Clinic
Country: Japan

Furukawa Clinic commissioned Mysa to design a completely new VIimage and signage system. The designers chose blue and red with regional features as main colour palette and designed a unique signage system, clear and legible.

古川内科诊所

古川内科诊所特别委托 Mysa 公司为其打造了全新的 VI 形象及标识系统。设计师选择了具有当地地域特色的蓝色和红色作为该系统的主色调,并设计了一套独特的标识符号,简洁易懂。

设计机构: Mysa 设计公司 设计师: 阿田成由 摄影: 山田贵史 委托方: 古川内科诊所 国家: 日本

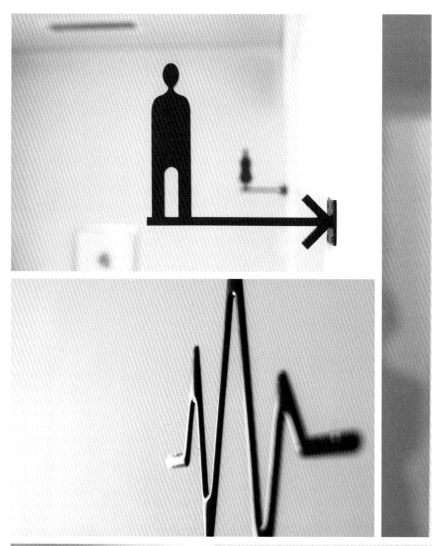

男性トイレ
MEN

検査室
LABORATORY

Dentestet 4 Kids Clinic

Design agency: Hamid- Nicola Katrib interior design studio
Designer: Hamid- Nicola Katrib
Photography: Alex Melenete
Country: Romania

The journey begins with the waiting and reception areas, which resemble a garden painted in pastel colours, so as to please the children, with flowers, birds, trees with fruits, all made in a simple, naive manner, perhaps inspired from the drawings made by little children. In the reception, the designer installed a couch for parents, who may watch images inside the examination rooms on screens, so as to see what happens during examinations, or watch the play zone, where children can enjoy a wii system, various games and coloured books, and during summer they also have the opportunity to play outside, in a little wooden house. After half an hour spent on the playground, they are called for examination in friendly rooms designed based on themes dedicated to children. On top of each dental chair there is a TV installed, so that the patients can watch cartoons throughout the visit to the dentist. The clinic has three examination rooms displayed in a distinct manner: the pirates' room, the puzzle room and the penguins' room. In addition, there is also a recovery room, where children may spend several minutes after interventions. Here they can colour books or play before leaving the clinic.

儿童牙科诊所

就诊的旅程从等候区和接待区开始，这里就像用柔和的色彩绘制的花园，用线条简单的花朵、小鸟、果树来取悦着孩子们。在接待处，设计师为家长设置了一个沙发，他们可以通过显示屏看到诊疗室内部的情景，也可以照看着在游戏区看书、玩游戏的孩子。在夏天，孩子们还可以在室外的小木屋里玩耍。在游乐场玩上半小时后，他们就可以进入专门为儿童设计的诊疗室进行治疗了。每个牙科治疗椅上方都有一台电视，患者可以边治疗边看卡通片。诊所的三间诊疗室以独特的方式呈现出来，分别是海盗屋、拼图屋和企鹅屋。此外，诊所还设有一间恢复室，孩子们可以在治疗后休息一下，看看图画书或玩一会儿游戏。

设计机构：Hamid- Nicola Katrib室内设计工作室
设计师：哈米德•尼古拉•卡特里伯 摄影：亚历克斯•梅勒尼特 国家：罗马尼亚

Clinic

Clinic

Clinic

ORL Clinic

Design agency: MALVI
Designer: Panos Voulgaris, Maria Malindretou Vika
Photography: Giorgio Papadopoulos
Client: Iakovos Arapakis, MD, PhD
Country: Greece

The client, a newly arrived surgeon, required a distinctive identity to match the memorable ORL Clinic. The approach consisted in associating the client (an ENT consultant/surgeon) to his practice and clinic, in order to create a professional, strong first impression, crucial for any newcomer in a saturated market. The three elements (person, specialty and studio) are connected through the use of the sound wave form as an interwoven visual element, which in turn can be used separately without compromising its effectiveness. The main waiting room's walls are covered with the Hippocratic oath, with some of its most important words standing out in lasercut acrylic glass.

ORL 诊所

作为一位新执业的外科医生，委托人要求设计师打造一个与诊所设计相匹配的独特形象。设计师决定将医生的执业情况与诊所联系起来，为人们留下专业、深刻的印象，这对新开业的诊所来说至关重要。人、专业、工作室三个元素通过声波图形交织起来，突出了诊所的专业感。主候诊室的墙壁上书写着"希波克拉底誓词"（医生誓约），其中重点词汇通过激光切割的有机玻璃获得了突出显示。

设计机构：MALVI 设计公司　设计师：帕诺斯·乌尔加里斯、玛利亚·马林德里托·维卡　摄影：乔吉奥·帕帕佐普洛斯　委托方：雅科沃斯·阿拉帕齐斯医生　国家：希腊

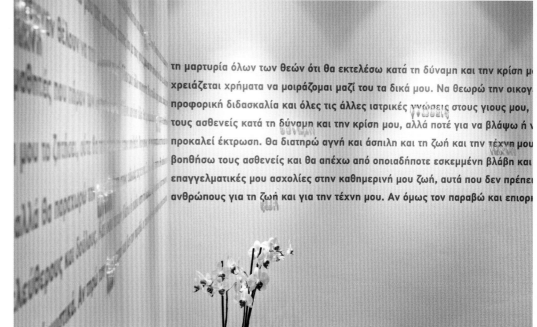

Clinic

Matsumoto Pediatric Dental Clinic

Design agency: TERADADESIGN
Photography: Yuki OMORI
Country: Japan

In the design for the Matsumoto Children's Dental Clinic, the designer utilised Accessory Sets. The designer was asked to create a space where children could enjoy a tensionless visit to the dentist, so the designer placed actual-sized Accessory Set animals to greet them before they walked in to the examination room. The designer also set playground equipment into the waiting room walls and designed it so it is viewable from the examination room, thus creating a space where children and parents can feel at ease.

松本小儿齿科医院

在松本小儿齿科医院的设计中，设计师选用了成套的装饰。设计师被要求打造一个让儿童不紧张的就医环境，所以他在诊疗室的入口处用实际大小的动物图案来欢迎他们。设计师还在候诊室的墙壁上设计了一套游戏装置，即使从诊疗室也能看到外面的游戏情景，为孩子和家长都营造了一个放松的空间。

设计机构：TERADADESIGN 设计公司 摄影：大森友希 国家：日本

Clinic

Dentist Kids Doc

Design agency: lalamoto Grafikdesign
Client: Dentist Kids Doc
Country: Germany

The reception of Dentist Kids Doc is designed as a cockpit. Through curved walls and aircraft windows, the examination rooms are directly visible. The design is based on warm colours, totally different from normal airports in cool colours. Graphic signs like those in real terminal guiding system provide direction guides for this 600-square-metre clinic. Where is pperating room, where is washroom, each door is clearly marked. The designers uses FF Netto as font for signs. Replaceable signs have considered children's vision height and trans-cultural intelligibility. The children are attracted by the design. At the check-in area, patients will get a "boarding check", and they will get a "suitcase" with small gifts when they are leaving.

儿童牙科诊所

这家儿童牙科诊所的前台设计以飞机驾驶室为主题，有弧度的墙体以及飞机窗的造型使人们可以直接看到诊室。整个设计的色彩基调为暖色，完全不同于一般冷色调的飞机场建筑。如同真正航站楼指引系统的图示符号为600平方米的诊所提供了方向指示。哪里是手术室，哪里是盥洗室，每一扇门都被清晰地做了标记。设计师将FF Netto定为图示符号字体，并加以扩充。易于更换的标识牌设计同时考虑到了儿童的视角高度和跨文化的可理解性。孩子们被医院的设计所吸引。在登记处病人会得到登机牌，离开时也会得到装有小礼物的旅行箱。

设计机构：lalamoto Grafikdesign设计公司 客户：儿童牙科诊所 国家：德国

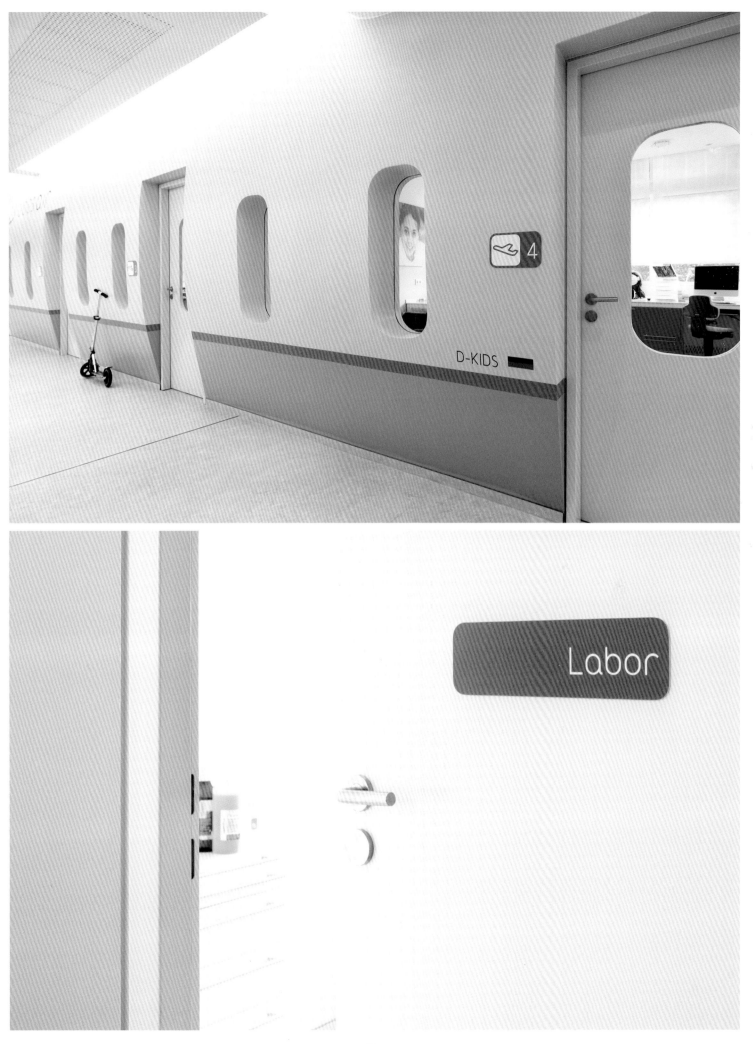

Clinic

Clinic

Clínica Dental Pons, Banyoles

Design agency: Estudi Arquitectura Interior Maite Prats
Designer: Maite Prats and Mariana Colmenero
Photography: Medir Cucurull
Client: Clínica Dental Pons
Country: Spain

The proposal for the Gemma Pons Dental Clínic interior emerges from the aim to create a welcoming, relaxing and warm atmosphere, but on the other hand the clinic design needs to be functional, away from traditional formal approaches. The Project was designed by Maite Prats and Mariana Colmenero from: Estudi d'Arquitectura Interior Maite Prats situated in the city of Banyoles, Girona, Spain.

巴尼奥莱斯庞斯牙科诊所

庞斯牙科诊所的室内设计目标是营造一种友好、放松、温馨的氛围。但是另一方面，诊所设计必须实用，不同于普通的空间设计。项目由来自西班牙巴尼奥莱斯市Maite Prats建筑与室内设计公司的马耶特·普拉斯和马里亚纳·科尔梅奈罗设计。

设计机构：Maite Prats建筑与室内设计公司 设计师：马耶特·普拉斯、马里亚纳·科尔梅奈罗 摄影：梅迪尔·库库鲁尔 委托方：庞斯牙科诊所 国家：西班牙

Clinic

Mónblanc Clínica Dental, Mataró

Design agency: Estudi Arquitectura Interior Maite Prats
Designer: Maite Prats and Mariana Colmenero
Photography: Medir Cucurull
Client: Mónblanc Clínica Dental
Country: Spain

The aim of Mónblanc Dental Clinic was maximised natural light received through a zenithal skylights, so the designers worked with transparent glass surfaces to play with natural light and gave the space character. The Project was designed by Maite Prats and Mariana Colmenero from: Estudi d'Arquitectura Interior Maite Prats situated in the city of Banyoles, Girona, Spain.

蒙布兰克牙科诊所

蒙布兰克牙科诊所的设计目标是最大化地利用天窗实现自然采光，所以设计师采用透明玻璃表面与自然光相配合，赋予了空间独特的风格。项目由来自西班牙巴尼奥莱斯市 Maite Prats 建筑与室内设计公司的马耶特·普拉斯和马里亚纳·科尔梅奈罗设计。

设计机构：Maite Prats 建筑与室内设计公司 设计师：马耶特·普拉斯、马里亚纳·科尔梅奈罗 摄影：梅迪尔·库库鲁尔 委托方：蒙布兰克牙科诊所 国家：西班牙

Clinic

Block722

Design agency: Sotiris Tsergas & Katja Margaritoglou
Designer: Eleni Meladaki, Katja Margaritoglou
Country: Greece

The starting commission of a reception redesign evolved into a unified refurbishment proposal of the whole pediatrician office, common and nursing spaces. A child's movement into the space becomes the starting point of the design process. The movements become spaces, areas, seatings and panels, creating a play-space for the young visitors. The originally divided space is unified, a process strengthened by the use of a single layer of linoleum floor. A polymorphous seating dominates the space, whereas all moving and dangerous furniture is replaced with steady constructions that allow free movement throughout the space. The archetypical shapes of the line and the circle coexist and form a welcoming space for the children, enriched with numerous toys and wall printings resized to a child's scale.

722 号诊所

对前台的重新设计最终延伸到了整个儿科医师办公室、公共区和护理区。设计以儿童在空间中的运动为出发点。这些运动变成了空间、区域、座椅和展板，为孩子们营造了一个游戏空间。单层油毡地板的使用让原本分开的空间被统一起来。一个多功能座椅在空间中起到了主导作用，所有移动而危险的家具都被稳固的结构所取代，保证了空间内的自由活动。典型的直线和圆形共同营造了一个欢迎儿童的空间，各种各样的玩具和反映儿童比例的墙画让空间变得更加丰富。

设计机构：Sotiris Tsergas & Katja Margaritoglou 设计公司 设计师：艾莱尼·麦拉达奇、卡特娅·马佳里托格鲁 国家：希腊

Clinic

Morinaga Maternity Clinic

Design agency: Mysa Co., Ltd Sign Design
Designer: Kosuke Yuge, Nariyoshi Kadota
Client: Morinaga Maternity Clinic
Country: Japan

Morinaga Maternity Clinic commissioned Mysa to create a new signage and identity system. With warmth and transparency as main concepts, the designers use a blooming flower as the clinic's logo. The four petals are green, purple, orange and yellowish green. Green stands for the "safety guarantee" that the clinic provides; purple for newly-born "healthy baby"; orange for women's "physical and psychological health"; yellowish green for the clinic's comfortable "environment and space". In the interior design, the designers create an art wall filled with "life flowers". Each floor is equipped with a unique colour, warm and romantic.

森永产科诊所

森永产科诊所委托Mysa公司为其打造了一套全新的标识及形象系统。设计师以温暖和透明为主要概念，并用绽放的花朵作为该诊所的标志，四片花瓣分别是绿色、紫色、橙色和黄绿色。绿色代表诊所为患者所提供的"安全保障"，紫色代表刚刚出生的"健康宝宝"，橙色代表女性的"身心健康"，而黄绿色则代表该诊所舒适的"环境和空间"。在室内，设计师特别打造了一面开满"生命花朵"的艺术墙，而每个楼层拥有独特的色彩，看上去温馨又浪漫。

设计机构：Mysa设计公司 设计师：羽多康介、角田成吉 委托方：森永产科诊所 国家：日本

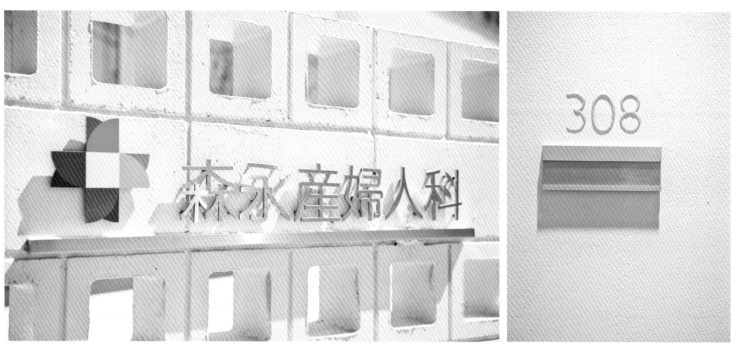

Clinic

Clinic

MURAKAMI
Dermatology Clinic

Design agency: Mysa Co., Ltd Sign Design
Designer: Mika Kato, Sayaka Yamaguchi
Client: MURAKAMI Dermatology Clinic
Country: Japan

MURAKAMI Dermatology Clinic commissioned Mysa to create a new signage and identity system for them. The whole design is based on black and white, simple and clear. Pink flowers are added as accents, highlighting the clinic's target group – women patients.

村上皮肤科诊所

村上皮肤科诊所特别委托 Mysa 为其打造了全新的标识及形象系统。整套设计以黑白为主，简单清晰，并以粉色的花朵作为点缀，突出了该诊所以女性患者为主要目标群体的特色。

设计机构：Mysa 设计公司 设计师：加藤美香、山口沙弥香 委托方：村上皮肤科诊所 国家：日本

Clinic

Clinic

Linden-Apotheke

Design agency: Ippolito Fleitz Group GmbH
Designer: Gunter Fleitz, Peter Ippolito, Fabian Greiner, Sascha Kipferling, Tim Lessmann
Photography: Zooey Braun
Client: Meike Raasch
Country: Germany

The Linden Apotheke is an old-established pharmacy in Ludwigsburg. It has chosen to specialise in naturopathic products and natural cosmetics in response to the growing pressure of competition in the pharmacy market. The remodeling of the pharmacy's interior serves to underscore and substantiate this positioning. The pharmacy's main focus should be communicated and made tangible within the space in a striking and compelling way – without creating a superficial, promotional style. At the same time, the task was to create additional possibilities of presenting shop merchandise in the shelving space behind the counter and the self-service area. Ippolito fleitz group was commissioned to undertake the interior remodeling of the space, revise the corporate design and develop a special giveaway for the grand reopening.

林登药房

林登药房是路德维希堡一座历史悠久的药房，主要经营自然疗法产品和天然化妆品，以此来应对竞争日益激烈的药房市场。药房室内设计的改造突出并强化了它的市场定位。药房的设计重点应放在显眼而令人信服的交流和沟通上，而不应给人以肤浅、营销的感觉。同时，项目还要求设计师在柜台后的货架和自助服务区进行有效的产品展示。Ippolito Fleitz Group对室内空间进行了重新布置，改良了企业形象设计并为药房的重新开张设计了特别的开幕活动。

设计机构 Ippolito Fleitz Group设计公司 设计师：甘特·福莱兹、彼得·依普利托、费边·格雷纳、萨沙·齐博菲林、蒂姆·莱斯曼 摄影：佐伊·布朗 委托方：迈克·拉施 国家：德国

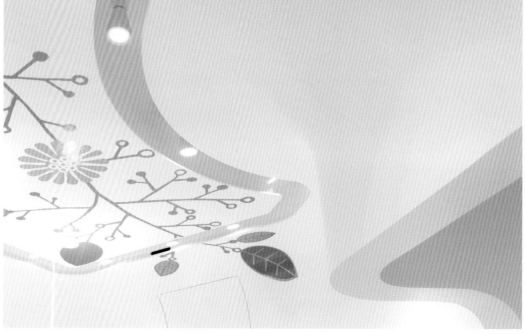

Pharmacy

Pharmacy

Whites Dispensary

Design agency: Studio Equator
Designer: Carlos Flores, Benjamin Fretard
Photography: Anne-Sophie Poirier
Client: Whites Dispensary
Country: Australia

In the design for Whites Dispensary, Studio Equator wanted to achieve a distinctly different look in an industry that is deeply rooted in a traditional and unchanging aesthetic. The aim was to design an interior and visual identity to increase the dispensary income whilst maintaining steady growth in the beauty, health and luxury goods sections, presented through an environment that emphasises personal service and a trustworthy brand. The ultimate goal is to reposition the brand to better service the market, and in doing so, change the way companies behave in an industry that hasn't changed in decades. With this refreshed concept, Studio Equator has allowed the visual identity to grow with the ability to change quickly and keep up with customer demands and challenges.

怀斯药房

在怀斯药房的设计中,Equator工作室希望打造一个与众不同的形象,打破药房根深蒂固的审美观。设计的目标是通过室内空间及视觉形象的提升来增加药房收入,通过突出个性服务和可信赖品牌的环境设计实现美容、健康和奢侈品类目销售额的稳定增长。设计的终极目标是重新进行品牌定位,更好地服务市场,最终改变数十年来公司在药房市场的经营方式。在这个概念下,Equator工作室的视觉形象设计能够快速变化,紧跟消费者的需求和市场的挑战。

设计机构:Equator工作室 设计师:卡洛斯·弗洛里斯、本杰明·弗里塔德 摄影:安·苏菲·波里尔 委托方:怀斯药房 国家:澳大利亚

Pharmacy

Pharmacy

Careland Pharmacy

Design agency: Sergio Mannino Studio (Store Design), Design Wajskol (Visual Identity)
Designer: Sergio Mannino Studio/Sergio Mannino, Martina Guandalini, Giulia Delpiano, Giulia Bortolotti, Design Wajskol/Jonathan Wajskol, Ying Fu, Janni Berger
Photography: Sergio Mannino Studio and Max Bolzonella
Country: USA

A dark shade of green is typically the colour associated with pharmacies, so the desigenrs pushed the colour to an extreme before it became so distant from what is commonly perceived as a pharmacy colour that people didn't recognise it anymore. The shade they picked is also a man-made colour, it is not a green associated with the idea of nature. Similar juxtapositions are found in the visual identity and the communications. Starting with the notion of "We take care seriously" the typography is a mix of a script, representing a friendly and warm approach, in contrast with an all caps sans serif font, a version of Futura, to emphasise the cleanliness, seriousness and precision of the operation.Virtually every pharmacy around the world uses the cross as an identifier so the desigenrs created their own version of the cross by drawing it by hand. While it has a friendly appearance, there is clearly a strong association with pharmacies and health.

凯兰德药房

深绿色是药房的传统色彩，因此设计师将这种色彩运用到了极致，以至于人们已经无法将其与药房联系起来。他们所选的色彩是一种人造色彩，并不是与自然相联系的绿色。类似的对立设计还呈现在视觉识别和交流系统上。以"我们认真对待"的理念为出发点，字体设计混合了手写体和全部大写的无衬线Futura字体，既显得友好而温暖，又突出了简洁感、严肃感和操作的精确度。几乎世界上所有的药房都用十字作为标识，因此设计师决定手写一个十字。手写的十字不仅看起来更友好，同时还呈现了与药学保健的紧密联系。

设计机构：Sergio Mannino工作室（店面设计）、Design Wajskol设计公司（视觉设计）设计师：Sergio Mannino工作室/塞尔吉奥·曼尼诺、玛蒂娜·古安达里尼、茱莉亚·德尔皮亚诺、茱莉亚·博尔托洛蒂；Design Wajskol设计公司/乔纳森·瓦基斯科尔、英·付、扬尼·伯杰 摄影：Sergio Mannino工作室、马克斯·波尔佐尼拉 国家：美国

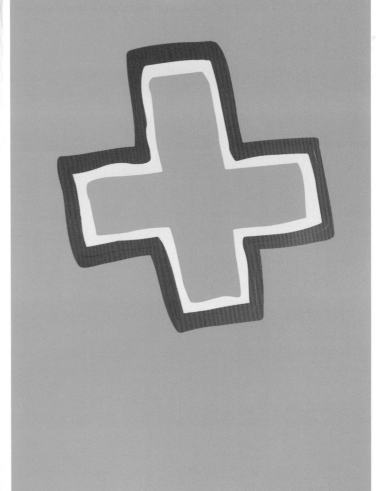

Pharmacy

PUR

Design agency: Bond Creative Agency
Designer: Aleksi Hautamäki, Toni Hurme, Janne Norokytö, Annika Peltoniemi, Jesper Bange, Lawrence Dorrington
Photography: Osmo Puuperä/All about everything
Client: Circlum Farmasia (Circlum Pharmacy)
Country: Finland

Pur is a new generation of wellness shops in Helsinki, bringing various aspects of healthy living together under one roof. Bond created a complete branding concept that covered everything from the brand identity to the shopfront design, website, photography, advertising and marketing collaterals. All this helps Pur to communicate its message of holistic wellbeing in an appealing, fun and informative way.

PUR 药房

PUR 是赫尔辛基的新一代保健药房，为顾客提供多样化的医疗保健产品。Bond 为其打造了全套品牌营销设计，覆盖了从品牌形象、店面设计、网站、摄影、广告和营销宣传品等方方面面。这些设计帮助 PUR 药房以动人、有趣的方式向消费者传递整体保健服务的信息。

设计机构：Bond 创意公司 设计师：阿列克西·豪塔玛基、托尼·赫尔姆、珍妮·诺拉奇特、安妮卡·佩尔托尼耶米、耶斯佩尔·班格、劳伦斯·多灵顿 摄影：奥斯莫·普佩拉 / All about everything 委托方：Circlum 药业 国家：芬兰

Pharmacy

Apótekarinn
Pharmaceutical

Design agency: GLÁMA-KÍM ARKITEKTS
Designer: Paddy Mills
Photography: © Martin Sammtleben
Client: Lyf & Heilsa
Country: Iceland

Apótekarinn was designed as a no frills brand. The applied narrative of medical packaging acts as a clear and appropriate visual metaphor for the design. The idea was to treat the counter like a piece of medical packaging. The dialogue between interior and graphic design processes worked well, delivering a clean, bold and strong identity for the brand.

阿普特卡林药房

阿普特卡林药房是一个廉价药房，不提供不必要的服务。药房的医药包装在视觉上体现了该品牌的设计特色：药房设计把柜台看成了一个药品包装。室内设计与图形设计相互作用，共同传递出简洁、大胆、强烈的品牌形象。

设计机构：GLÁMA-KÍM建筑事务所 设计师：帕蒂·米尔斯 摄影：马丁·萨姆特尔本 委托方：Lyf & Heilsa公司 国家：冰岛

Pharmacy

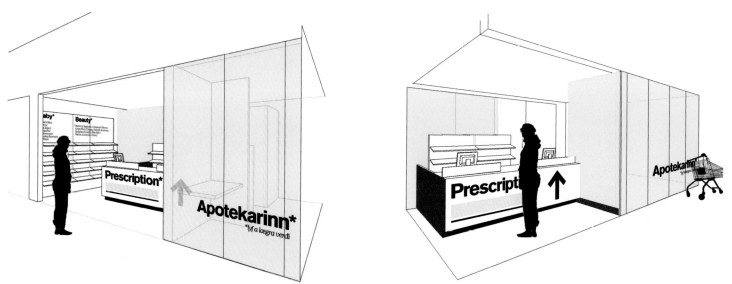

Pharmacy

索引

Alex Daye Design
180

Anne Gordon Design
100

Bond Creative Agency
232

clasebcn design
094

Dan Pearlman
052

Dotdash
062, 170

emmanuelle moureaux
182

Estefanía Belén Tinto & Victoria Alfieri
106

Estudi Arquitectura Interior Maite Prats
206, 208

Estudi Conrad Torras
186

Estudi E. Torres Pujol
128

GLÁMA-KÍM ARKITEKTS
234

Gourdin & Müller
168

Greco Design
068

Gresham, Smith & Partners
074, 116, 144, 150, 156

Hamid- Nicola Katrib interior design studio
192

ID/Lab
086, 092, 172, 176

Image Resource Group
140

Ippolito Fleitz Group GmbH
132, 220

Jennings design studio
166

lalamoto Grafikdesign
202

Leaf Design Pvt. Ltd.
040, 082

MALVI
198

MCA Manoel Coelho Arquitetura & Design
096

Morag Myerscough, Donna Wilson, Miller Goodman,
Tord Boontje, Chris Haughton, Ella Doran
044

Mysa Co., Ltd Sign Design
114, 190, 212, 216

PAUZARQ arquitectos
184

Perkins Eastman
072

QUA DESIGN style
158

RTKL Associates Inc.
122, 162

SCENO Environmental Graphic Design
080

Sergio Mannino Studio, Design Wajskol
228

SMITH UK LTD
034

Sotiris Tsergas & Katja Margaritoglou
210

Stanley Beaman & Sears
006, 010, 014, 018, 024, 028

Studio Equator
224

studio gd
118

StudioSigno
056

TERADADESIGN
200

umsinn.com
134

图书在版编目（CIP）数据

医疗导视. 2 /（美）哈丁编；常文心译. -- 沈阳:辽宁科学技术出版社，2015.5
ISBN 978-7-5381-9192-9

Ⅰ．①医… Ⅱ．①哈… ②常… Ⅲ．①医院－导视设计－世界－现代－图集 Ⅳ．①TU246.1-64

中国版本图书馆CIP数据核字(2015)第071414号

出版发行：辽宁科学技术出版社
　　（地址：沈阳市和平区十一纬路29号　邮编：110003）
印　刷　者：利丰雅高印刷（深圳）有限公司
经　销　者：各地新华书店
幅面尺寸：215mm×285mm
印　　张：15
插　　页：4
字　　数：50 千字
印　　数：1～1500
出版时间：2015年 6 月第 1 版
印刷时间：2015年 6 月第 1 次印刷
责任编辑：周　洁
封面设计：关木子
版式设计：关木子
责任校对：周　文
书　　号：ISBN 978-7-5381-9192-9
定　　价：258.00元

联系电话：024-23284360
邮购热线：024-23284502
http://www.lnkj.com.cn